Green Equilibrium

GREEN EQUILIBRIUM

The Vital Balance *of* Humans & Nature

CHRISTOPHER WILLS

With Photographs by the Author

OXFORD

UNIVERSITY PRESS

OXFORD
UNIVERSITY PRESS

Great Clarendon Street, Oxford, OX2 6DP,
United Kingdom

Oxford University Press is a department of the University of Oxford.
It furthers the University's objective of excellence in research, scholarship,
and education by publishing worldwide. Oxford is a registered trade mark of
Oxford University Press in the UK and in certain other countries

© Christopher Wills 2013

The moral rights of the author have been asserted

First Edition published in 2013

Impression: 1

British Library Cataloguing in Publication Data

Data available

Library of Congress Cataloging in Publication Data

Data available

ISBN 978–0–19–964570–1

Printed in Great Britain by
Clays Ltd, St Ives plc

LOST.

I lost a world the other day.
Has anybody found?
You'll know it by the row of stars
Around its forehead bound.

A rich man might not notice it;
Yet to my frugal eye
Of more esteem than ducats.
Oh, find it, sir, for me!

Emily Dickinson

Learn hence for ancient rules a just esteem;
To copy Nature is to copy them.

Alexander Pope, *Essay on Criticism*

To Liz,
my indispensable partner and the shrewdest of
observers of the human world.

Contents

List of Plates

List of Illustrations

Introduction

Zebras gambol and wildebeest graze in Tanzania's flower-filled Ngorongoro Crater. The many species that live in the crater have reached a precarious ecological and evolutionary balance, a green equilibrium.

Tanzania's Ngorongoro Crater, with an area of 260 square kilometers, is a caldera, the collapsed remains of a volcano that was once as large as Mount Kilimanjaro. The floor of the caldera is fed by groundwater that seeps down from the rim, so that water is always available for thriving populations of predator and prey animals.

When we traveled around the crater in the spring of 2008, it was ablaze with wildflowers. Herds of zebras and blue wildebeest grazed on the lush grass. Lions, cheetahs, and leopards stalked them, while graceful caracals hunted smaller prey. Jackals splashed into the shallows of Lake Magadi and returned proudly with struggling flamingoes in their jaws. A nearsighted rhino blundered like Mr. Magoo across the grassland of the crater floor. When the rhino accidentally approached a pride of lions at a wildebeest kill, they panicked and fled.

Ngorongoro's ecosystem is still largely intact, having survived intense tourism and the conversion of much of the surrounding area into grazing land. But it is still subject to the inexorable laws of evolution and ecology.

Although human manipulations of the crater's environment seemed at first to be small, they have had escalating consequences. During the quarter of a century prior to 2001, the park's rangers suppressed fires in the crater because lush and plentiful grass made the tourists happy. Ticks found shelter in the thick grasses and multiplied (Figure 1). A variety of tick-borne diseases soon began to threaten many of the crater's animal populations. Big cats were especially vulnerable, because their reproductive rates could not keep up with the deaths from disease. The numbers of lions, cheetahs, and other predators in the crater fell, and some populations of their prey began to increase unsustainably. Clearly, the crater's ecosystem was losing its balance.

Figure 1 These ear-ticks, clinging to a grass stem in Ngorongoro Crater, carry many animal diseases that form an essential part of the crater's green equilibria. Photograph courtesy of Winston Trollope.

In 2002 Tanzanian Veterinary Research Officer Robert Fyumagwa teamed with Winston and Lynne Trollope, a husband and wife team of grassland ecologists from South Africa, to attack this problem. Under Fyumagwa's direction, rangers began controlled burns of the crater grasslands.

Their intervention worked spectacularly. The number of ticks dropped, and populations of the big cats and other predators recovered [1]. The controlled burns did not rid the grasslands of the ticks and the diseases that they carried. Instead, they reduced the numbers of ticks. The result was a restoration of a delicate equilibrium that involved some of this ecosystem's essential components: the grasses, the periodic fires that swept through them, the populations of predator and prey animals, the ticks, and the diseases that the ticks carried. All these components, even the deadly ones, were essential to the balance. None could predominate without throwing the system out of equilibrium.

In this book we will examine the rules that govern such interactions among species. I call these ecological balancing acts *green equilibria,* because they keep our world vibrant, verdant, and ecologically intact. We will also explore how the world's green equilibria have shaped the evolution and history of our own species. And finally, and most importantly, we will see how evolutionary and ecological pressures similar to those that have produced these green equilibria have actually given our species the ability to undo the damage we are doing to our planet.

Green equilibria result from a wide variety of environmental selective pressures that act to maintain the diversity of species in ecosystems. The selective pressures have numerous origins, but many of them share one feature in common: their strength, and even their overall direction, varies as the species that they act on rise and fall in numbers and as the ecosystem itself changes. As the relative frequencies of species in ecosystems change, their interactions change as well. As we begin to explore balanced and unbalanced ecosystems in detail, we will find that such *frequency-dependent* selective mechanisms are often contributors to the ecosystems' survival.*

The huge numbers of ticks that appeared in the Ngorongoro grasslands served as a clear early warning to the park's naturalists that its green equilibrium was losing its balance. Other, less obvious, factors contribute to the green equilibria of every ecosystem. One important type of frequency-dependent balance happens at the level of genes, resulting in a kind of genetic green equilibrium. At the genetic level the frequencies involved are the relative frequencies of genes in a gene pool, rather than the relative numbers of different species in an ecosystem.

Each species in an ecosystem possesses a *gene pool*, the sum of all the genes that are carried by all the members of the species. As a result of mutations, each species' gene pool has become filled with different forms of many of its genes. Geneticists call these different forms *alleles* (short for *allelomorphs*, "other forms").

Selective pressures can change a gene pool by increasing or decreasing the relative numbers of these alleles. New alleles can also flood into a gene pool from different populations, and even from allied species—as we will see when we examine our own evolutionary history.

Some alleles sweep through populations and transform them, but many others are maintained in the populations' gene pools by a balance of frequency-dependent selective forces. It is such balances that establish genetic green equilibria.

None of these green equilibria, ecological or genetic, are permanent. When an ecosystem's environment changes, both the ecological equilibria and the genetic equilibria of each of its species must shift and change as well. But many ecosystems that have had the time to reach equilibrium tend to have high ecological diversity and high within-species genetic diversity. They can draw on this diversity when the environment changes, increasing the chance that some members of the ecosystem will survive.

---

* Ecologists distinguish between such *frequency*-dependent effects, which arise as the relative numbers of different interacting species change, and *density*-dependent effects, which happen when a species' numbers change and it uses more or less of its resources as a result. Density-dependent effects could take place in an ecosystem that has only one species, while frequency-dependent effects measure the interactions among species.

The human species has also been shaped by these genetic and ecological pressures. On the positive side, our gene pool has accumulated a huge collection of genetic alleles that have helped to give us the greatest range of intellectual and physical abilities of any species that has ever lived on our planet. Robert Fyumagwa and his rangers drew on these powers to understand and maintain the Ngorongoro Crater's green equilibrium.

On the negative side, we have used these remarkable abilities of ours to push many of the world's green equilibria out of balance. Since the agricultural revolution began about ten thousand years ago, we have utterly changed almost half of the land surface that could conceivably be used to raise crops [2]. In the process we have caused entire ecosystems to disappear, sometimes without a trace. We have changed other ecosystems so drastically that their current state is unsustainable.

A recent flowering of technology has enabled our population to multiply to seven billion people and counting. The result has been widespread ecological damage. But it is also remarkable, and a tribute to our new technology, that a larger fraction of our own species than ever before is able to live long, productive lives. In the years 1965–70, median life expectancy for the more than two billion people living in Asia was 54 years. Thirty-five years later, in the period 2000–05, median life expectancy throughout that vast continent had risen to 67 years. There has been an astonishing increase of nine months in average life expectancy for every two years of calendar time during this period [3]. This incredible change for the better has taken place in spite of wars, and even despite the terrible genocide in Cambodia that killed more than three million people in the late 1970s.

During this same period there were similar but less extreme increases in life expectancy elsewhere in the world. This progress has been sparked by improvements in agriculture, the spread of education and public health, improved nutrition, better food transport, and most importantly the empowerment of women.

Our behaviors have changed for the better in many ways. Steven Pinker, in his magisterial book *The Better Angels of Our Nature*, has recently documented how the spread of civilization and societal interconnectedness, along with our increased emphasis on the worth of the individual, have had the effect of reducing levels of violence in every part of the world [4].

But these unparalleled successes for our own species have come at the cost of unsustainable overexploitation of the world's resources. There are three possible paths that we can follow as we confront this threat to our very existence as a species.

First, we can blindly continue to exploit and damage all the world's ecosystems, regardless of how irreplaceable they are or how many other species we force into extinction. This is plainly suicidal.

Second, we can continue our merciless exploitation of much of the planet, but also set aside some small areas such as Ngorongoro to serve as refuges for the world's wildlife. This course requires that we become good and effective stewards of the refuges. As we will see, our track record for such stewardship is decidedly mixed. There are real dangers if we follow this path exclusively.

Third, we can modify our behavior by applying what we are learning about the principles of ecology and about the evolution of different species, including our own. Until the human population eventually contracts to more manageable numbers, our species will continue to be the Earth's ruling predator. But there is no reason why, even during the next overcrowded century, we cannot maintain a large part of the world in a healthy balance. We need to learn from the natural world how to be cleverer and less exploitive predators. We must learn how to understand and harness the powerful evolutionary processes that have enabled other animals and plants to live together in green equilibria.

It is this third path that we will explore here. Before humans, no single species has had such an impact on the planet's ecosystems. But, before humans, no species was intelligent enough to consciously modify its behavior. Although we have been sufficiently intelligent to do this for hundreds of thousands of years, we can plead the excuse of ignorance. But now, for the first time, we can draw on our growing pool of knowledge and on our new communication technologies to fully utilize our collective intelligence as a species. As we will see when we explore the evolution of the human brain towards the end of the book, we now have no more excuses.

As we explore the ramifications of this third path, we will travel to ecosystems in California, Guyana, Brazil, the Philippines, the central Pacific, Thailand, New Guinea, Nepal, Bhutan, and more. Some of these places are relatively pristine, others are threatened, and a few are recovering from human damage.

During our travels we will meet some of the many ecologists who have gathered data about frequency-dependent interactions. And we will meet local people who are being trained by these scientists in scientific methods, giving them for the first time the tools that they need to understand the environment that has nurtured them.

To understand these ecosystems in depth, we must also explore their history. For example, the geological stability of the ancient Guiana Shield of northern

South America helps to explain why so many ancient lineages of animals and plants have survived there. And the geological upheaval that resulted from the collision between the Indian tectonic plate and the massive South Asian plate has produced, not only the massive foothills and towering peaks of the Himalayas, but also some of the world's newest, most extreme, and most fragile ecosystems.

As we have changed these ecosystems, the ecosystems have changed us. Recent genetic studies have reinforced Charles Darwin's conviction that we have been shaped by the same ecological and evolutionary pressures that have acted on other species.

Our ecosystems have shaped us in many ways. New genetic data are revealing how the peoples of the Himalayas and Tibet have adapted to life at high elevations. At the same time, other groups of people who live only a hundred kilometers to the south of the Himalayas have adapted in different but equally dramatic ways to their tropical lowlands. And some of these lowland people are currently migrating into the mountains, pushing to the limit their ability to survive under these extreme conditions.

When the descendants of the first modern human migrants out of Africa arrived in New Guinea fifty thousand years ago, at the end of a journey that covered the width of Asia, they found ecosystems ranging from disease-ridden lowland tropical forests to sparkling alpine meadows. The result has been a fragmentation and a rich diversification of human societies on the island that could only have taken place in such a complex environment.

We will also explore how these people brought with them the ingredients of a new kind of genetic equilibrium. Along with the genes that they carried from their ancestors in Africa, they had accumulated genes from at least two groups of people who preceded them, the Neanderthals and the mysterious Denisovans. This complex genetic heritage of the people of New Guinea, such as Chief Wilem of Wasilimo Village in New Guinea's western highlands (Plate 1), has helped them to adapt to their equally complex ecological world. Their remarkable genetic heritage will continue to aid them as they and their children confront the world of the future.

In spite of such human–ecosystem coadaptations, long before the time of recorded history our negative effects on the planet have been far out of proportion to our numbers. The story of the colonization of North and South America by hunter-gatherers from Asia, with its accompanying wave of extinctions of about a third of the large animal species on these continents, shows vividly the disconnect between our potential as a species and the history of what we have done to the world.

Can we learn from mistakes like these, mistakes that result from our ignorance? We have the equipment. We are now beginning to understand, at the genetic level, how we have acquired such large and versatile brains. We will soon be able to trace how those brains have evolved, and to explore the contribution of genetic green equilibria that have helped to make us such a remarkable species. We will discover that our collective intellectual history has resulted in a kind of mental collective, a diverse rainforest of the mind.

Finally we will turn to some cases in which entire groups of peoples are using those amazing brains to make the leap into modernity and—perhaps—to restore the green equilibria of their environments. The stories of the Himalayan countries of Bhutan and Nepal are complex and challenging but laced with hope. And in the process we will discover another dimension to such stories, one that is unique to our species. This extra dimension will help us all as we begin the task of healing our planet.

At the end of the book I will take you to a place where human art and the art of nature are literally intertwined. This magical spot serves as a metaphor for the third way that we must follow if our species, and our planet, are to survive.

About the photographs

As in my recent book *The Darwinian Tourist*, all the creatures in these pictures were living in the wild when I photographed them. I have made some adjustments of color balance and contrast, along with some burning and dodging that have allowed me to emphasize subjects that are on a confusing background. I have also retouched the underwater pictures slightly, in order to erase the tiny creatures that show up as bright specks in even the clearest water. Otherwise, nothing has been added to, or removed from, the pictures.

Some of the photographs are in special color sections, and descriptions of them are given in the text.

Acknowledgments

As my wife and I have traveled the world, both in pursuit of rainforest research projects and to satisfy our curiosity about how the living world works, we have been aided by a remarkable group of scientists, guides, and friends who have made our way easier. Many of them have prevented me from making mistakes of fact, and have warned me about marching over the edge of what is known into realms of speculation. I have done my best to correct the errors, and any that remain are entirely traceable to my own obtuseness. These most helpful people include Orlando Arciaga, Steve Cande, Catherine Demesa, Pascal Gagneux, Thomas Higham, Lyndal Laughrin, Lucio Luzzatto, James Mallet, Emma Marchant, Kristen Marhaver, Donna Moore, Dick Norris, Franck Polleux, Vojtech Novotny, David Reich, Winston Trollope, George Weiblen, and John West. Friends, relatives, and colleagues around the world have helped in uncounted ways, among them Buddha Basnyat, James O'Donoghue, Hopi Hoekstra, Elsa Cleland, Todd Tyler, Giovanni Paternostro, Barbara Scholz, Mark Strickland, Anne-Marie Wills, and Liz Fong Wills. I am most grateful to scientists and parataxonomists working in the world's most remote regions who spent time explaining what they were doing: John Auga, Mark Broomfield, Phil Butterill, Sarayudh Bunajachewan, Rick Condit, Chris Dahl, Kyle Harms, Don Hartley, Milan Janda, Samuel Joseph, Fernando Li, Samuel Muwi, Kenneth Molem, Chandra Sambhu, and Nason Semi. I cannot acknowledge all the many guides and hosts who were so essential for our travels, but I am especially indebted to Liza Bardonado-Dahl, Mark Castillo, Nanda Chai, Lerin Facão de Arruda, Eduardo de Arruda, Andrea and Salvador de Caires, Ashley Holland, Kate McCurdy, Diane McTurk and Will Weber, along with Bhutan guides Karma Dorji and Sonam Choki and Nepal guides Raju Chaudhari, Raj Kumar Mahato, Kama Thinba Sherpa, and Pasong Sherpa. Barbara Jenkins-Lee and Katie Stoyka, our indefatigable travel agents, made our journeys astonishingly problem-free. Special thanks to my editor Latha Menon, who kept me focused and lured me back from those rococo byways that so tempt an author. Phil Henderson and Kate Farquhar-Thomson at Oxford University Press also provided valuable input. This book has been my most complex enterprise to date, and it would never have happened without all of you!

CHAPTER ONE

How Ecosystems Work

About 80 kilometers north of Santa Barbara, nestled in California's coastal mountain range, is a cattle ranch that is now known as the Sedgwick Reserve. Although the ranch had been grazed heavily for two centuries, since 1997 it has been largely protected from human activity and is slowly changing to a wilder state.

The Reserve includes over 2,500 hectares of beautiful coastal range land (Figure 2). Because of the terms of the conservation agreement a few cattle still roam a hundred of these hectares, but most of the ranch is now free to change to a different kind of ecosystem.

In the summer of 2010 I drove up to the Reserve through miles of horse-rearing country. I wanted to explore what happens when such ecosystems, in California and elsewhere, have been protected from grazing, cultivation, or fishing.

The Reserve includes two parallel narrow valleys, each spanning a mix of forest and grassland. It is possible to drive up each of the valleys along dirt roads. The roads both climb about 500 meters, passing through a series of small ecosystems that range from grassland to oak forest to pine forest. Eventually the roads end at a fire break that separates the Reserve from the range of coastal mountains behind it.

Mule deer can now run freely through the summer grass of the Sedgwick Reserve (Figure 3). Surveys are picking up increasing signs of coyotes, black bears, foxes, mountain lions, bobcats, badgers, and weasels. Wild turkeys are often seen, and there are several nests of golden and bald eagles. The local populations of many other birds seem to be increasing. But some of the once-iconic animals of California's coastal ecosystem—grizzlies, elk, and pronghorn antelope—are unlikely to recolonize the Reserve. They have been hunted to local extinction over too wide an area in southern California.

The Sedgwick Reserve has become a magical place, vibrating with new life. But what are the rules that such ecosystems follow when they are released from

Figure 2   Scrub and live oak trees dot this quintessentially Californian landscape. Here, at the Sedgwick Ranch in the coastal foothills north of Santa Barbara, the hills are covered with golden wild oats, and the tracks of cattle are still visible. Because most of the ranch is now protected from grazing, it is starting to revert to a wilder state. But, because this protected land is surrounded by human-impacted ecosystems, its possible futures will be governed by both human and natural influences. The ranchland is clearly changing, but can it change to a truly wild state in a world that is becoming less wild?

human influence? When do they revert to some previous state, and when do they morph into something new?

The Structure of Ecosystems

Ecology, from the Greek for "knowledge" and "household," is the science that seeks to understand the households that the world provides for us and all other living species. These households are our planet's ecosystems. But just what is an ecosystem?

In 1971 the ecologist Eugene Odum proposed a rigorous definition of this basic ecological concept:

Any unit that includes all of the organisms (i.e.: the "community") in a given area interacting with the physical environment so that a flow of energy leads to clearly defined

Figure 3 These California mule deer, *Odocoileus hemionus*, are now free to run through the Sedgwick Reserve. They ignore the old fence that you can see behind them, and are not disturbed by ecological research enclosures like the one in the foreground.

trophic structure, biotic diversity, and material cycles (i.e.: exchange of materials between living and nonliving parts) within the system is an ecosystem [1].

This is a forbidding and jaggedly technical chunk of prose, made even worse by its Germanic sentence structure. Let us try to deconstruct it.

Odum says that an ecosystem is made up of all the organisms that occupy a given region of the biosphere, the thin layer of the Earth that can support life. He points out that the organisms are affected by that region's physical characteristics, and that the organisms are grouped into a "trophic [nutritional] structure" because of the ecosystem's energy flow.

The sun is the primary source of this energy flow. In every ecosystem the *lowest trophic level* is occupied by green plants that obtain their energy from the sun. In terrestrial ecosystems like the Sedgwick Reserve these *primary producers* are the familiar grasses, bushes, and trees, and in aquatic ecosystems the lowest level is made up of innumerable tiny one-celled plants that float near the water's surface where they can be bathed by sunlight.[1]

3

These plants can capture about ten percent of solar energy through photo-synthesis. The other ninety percent is lost, mostly as heat.

In a typical terrestrial ecosystem like the Sedgwick Reserve, some of the plants' energy is captured in turn by *primary consumers*, the wide variety of birds, mammals, and insects that dine directly on the plants. But much plant material is not eaten. Animal primary consumers can capture only about ten percent of their energy, which means that they are only able to utilize one percent of the original energy from the sun.

About ten percent of this one percent is then captured by the first of a series of additional trophic layers of predators that eat these vegetarian browsers, graz-ers, and nibblers. At the Sedgwick Reserve these *secondary consumers* include small snakes that eat field mice, and birds such as hermit warblers that dine on insects. The energy stored in the bodies of these predators, three steps away from the energy of the sun, represents about a tenth of one percent (ten percent times ten percent times ten percent) of the solar energy that was originally received by the ecosystem (Figure 4A).

The summed mass (*total biomass*) of the plants or animals at each of these trophic levels is regulated by the amount of energy that the organisms in that level have been able to obtain from the level below them. At each stage, that summed biomass is far smaller than the biomass of the level below.

Large ecosystems are able to support as many as five or six trophic levels. The higher levels are made up of predators that eat the predators that occupy the next level down—hawks that can eat snakes and small birds, or mountain lions that can eat coyotes, for example. The predators of each succeeding level capture tinier and tinier fractions of the original energy flow (Figure 4B).

On the Sedgwick Ranch, until recently, herds of grazing cattle knocked this "ten percent rule" out of balance. Their trampling hooves, coupled with the periodic fires that swept through the hills, kept bushes and saplings from invading the grasslands. Each year the grazed and burned-over land quickly sprouted water-hungry grasses, providing an ideal environment for the cattle. This interrupted the flow of energy that had once supported native animals at higher trophic levels.

Humans had changed the ranch from a complex web of ecological interac-tions into a system designed to produce cow-flesh. This artificial ecosystem would have been unsustainable in the long term. The monoculture of grass was depleting the soil of nutrients. Each dry season the cattle and the fires scalped the hills. Winter storms eroded the bald landscape, removing topsoil and silting up the streams. Eventually, the ranch would have been unable to support cattle.

Figure 4A Geologist Todd Tyler of California State University at Long Beach snapped this picture of a mountain lion, *Puma concolor*, in early 2010 at his study site near the southern edge of the Sedgwick Reserve. Top predators such as mountain lions and bald eagles are becoming more common in the area.

(b)

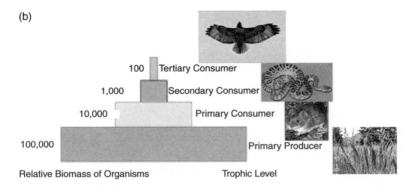

Figure 4B The total biomass of organisms decreases rapidly with each succeeding step up in the trophic "pyramid." The diagram is a simplification, because many species play a role in more than one trophic level. For example, a red-tailed hawk may eat rattlesnakes and mice.

When the ranch was turned into a reserve, most of these destructive processes were halted. But the intensive grazing that took place during the last two centuries is only the latest modification of an ecosystem that had already undergone vast human-caused changes. To understand the present ecosystem of the Reserve and its possible futures, we must understand these changes.

Sedgwick's Long History

When the first human migrants from central Asia reached California, probably about 13,000 years ago, they found mountain and valley ecosystems that were almost unimaginably rich and varied. Native American horses, long-horned bison, mammoths, ground sloths, mastodons, and camels roamed California's hills and its great Central Valley in vast numbers. All of these grazers and browsers, along with the lions, saber tooth cats, and dire wolves that preyed on them, vanished at about the same time as the first human hunters appeared on the scene.

In Chapter Ten we will trace in detail the story of how these early peoples, through their hunting and the vast fires that they ignited, helped to transform the ecosystems of the Americas into a less species-rich world. But even after this transformation huge numbers of potential prey species in California still remained, including large herds of elk, deer, and antelope. These helped to support a substantial human population, which gradually diversified into a kaleidoscope of tribes.

The first Spanish explorers who sailed up the California coast laid claim to an enormous territory that they called Alta (Upper) California. They explored only a tiny part of the lands that they so blithely appropriated, and had no idea of their diversity. Even what is now Northern California, a small part of this new province, supported five hundred tribes who spoke fifty dialects of six major language groups [2]. The ecosystem changes that had been triggered by the arrival of their distant ancestors had faded beyond living memory. The tribes lived in new, less diverse, ecosystems that had nonetheless managed to reach new tentative equilibria.

European settlers and missionaries began to arrive in California in substantial numbers during the last third of the eighteenth century. They brought devastating diseases that decimated the native peoples. And they also brought new grazing animals and crop plants. Together, these events triggered further overwhelming changes in these already profoundly altered ecosystems.

The ecosystems of present-day California have global origins, dominated by invasive plant species that have been introduced by immigrants from around the world. A 1998 state-wide survey found 1,075 introduced plant species [3]. During the last two hundred and fifty years, the regions where native animals and plants are still able to survive have been reduced in size and fragmented. Domestic cats are now the top predators in many parts of California.

Two hundred and fifty years represents three long human lifetimes. Today's Californians have grown up in a world that has little resemblance even to the profoundly altered ecosystems that greeted the first Europeans.

The history of the Sedgwick Ranch itself is a microcosm of California's complicated history. When Mission Santa Ynez was founded in 1804, it claimed a great swathe of land that included the site of the future ranch. About 2,000 Chumash, Tulereño, and other tribes were living on the mission lands, but they were being decimated by introduced diseases at the same time as the poor and unschooled Spanish and *mestizo* colonists known as the *Californios* began to arrive.

After the 1821 Mexican revolt that freed California from Spanish rule, the new Mexican government gave some of these *Californio* settlers immense parcels of land near the coast. The government also secularized the missions in 1834, and much of their land was given away. Somehow, Santa Ynez managed to hang onto its lands for years afterwards.

Then settlers from the United States began to arrive. When Alta California was ceded to the U.S. in 1848, at the end of the Mexican–American war, only a few of the *Californios* managed to retain their ranches. Many of the ranches went into foreclosure and ended up in the hands of the Americans. Most of the remaining *Californio* property holders lost their lands after a huge flood in 1861, which was followed in 1863 by a disastrous drought.

The land that was to become the Sedgwick Ranch passed through several hands during this period. Pio Pico, the *de facto* governor of California before the Mexican–American War, deeded a parcel of land including the ranch to an army veteran, Octaviano Gutierrez, in 1845. After the great flood and drought, Gutierrez was forced to sell the land to a group of Yankees. His widow tried unsuccessfully to get it back. Over the next eighty years, the land's ownership dissolved into a welter of lawsuits and confusing transactions [4].

In 1951 the ranch was bought from its most recent American owner by a branch of an old and distinguished Massachusetts family, the Sedgwicks. This California branch is chiefly remembered for a daughter, Edie, who became one of Andy Warhol's "muses" and starred in many of his films. Edie's fifteen minutes of fame ended with her death from a drug overdose in 1971. A sad little tribute to Edie's life, in the form of a series of portraits by Warhol, now hangs on a wall of the Reserve's ranch house.

When Edie's father, Francis, died in 1988, the Santa Barbara Land Trust bought out the family's remaining interest. The entire ranch is now administered by the University of California at Santa Barbara, and has become part of the university's Nature Reserve system. University ecologists now use the Reserve to carry out a wide variety of experiments.

The Golden Hills

The Sedgwick Reserve of today would be unrecognizable to the Chumash and Tulereño tribespeople who had lived there before the arrival of the Spanish.

The land that they knew has been transformed by the invaders. It has turned to gold.

California is, of course, the Golden State. One of the origins of these golden nicknames is certainly the discovery of gold that catapulted California onto the world stage in 1849. To the Cantonese laborers, who built the railroads and worked the gold fields, California was *Gum Shan*—"Gold Mountain." Another likely origin can be traced to the fields of orange poppies that bloom in the state's remaining grasslands each spring. But there is a third source for this golden image. Everywhere you travel along the coastal ranges and among the foothills of the Sierras during the summer and fall, the grass-covered hills glow with a color that resembles the golden landscape of Spain's hill country.

The resemblance is not a coincidence. Much of the golden color comes from *Avena fatua*, an annual wild oat that is native to the Mediterranean and North Africa. This species was introduced to eastern North America by some of the first European colonists, more than three centuries ago. It reached California a hundred years later.

By a happy chance, because adobe bricks are made with a mixture of straw and mud, archeologists have been able to document with precision the rapid spread of this grass. There are no signs of *Avena* straw in the bricks that Father Junipero Serra's native laborers used when they built the first California missions in the 1760s and 1770s. But by the early nineteenth century this wild oat straw had become a ubiquitous component of adobe bricks throughout California [5], [6].

What color were California's hills before the arrival of *Avena fatua* and its close relative *A. barbata*? We cannot be sure, but they were probably covered with reddish-brown native perennial oatgrasses and fescues. The valleys between the hills supported clumps of green and brown reedgrasses.

Avena fatua and other introduced annual grasses have won out over the native perennials because they grow more quickly and are superb at grabbing all the available water. Russian experimenters grew a single *A. fatua* plant for 80 days in the absence of competitors. Then they added up the lengths of all its mass of roots. The total sum was an astonishing 80 kilometers [7]!

These invading grasses triggered further changes. The two valleys in the Reserve are separated by a little range of golden, *Avena*-covered hills. As I climbed up their slopes I stirred up clouds of tiny grasshoppers that leaped ahead of me in an ornithopteran tidal wave. These were immature forms of a native grasshopper with the ominous name of *Melanoplus devastator*. Plate 2 shows that they matched the color and size of the little *Avena* florets perfectly.

M. *devastator* lives up to its name. It has been devouring farmers' fields at least since the time of the early Spanish settlers. The worst outbreaks, in the 1960s, affected more than one and a half million acres. The grasshoppers consumed vast areas of grassland, along with the leaves of tens of thousands of hectares of grape vines and fruit trees. They still swarm in the Sedgwick Reserve's grasslands, but their malign influence is diminishing as the grasslands begin to alter and become more complex.

You Can't Go Home Again

The grass and grasshoppers on the hills of the Sedgwick Reserve make up a strangely simplified ecosystem, a mere faint echo of the diversity of the past. But now, the grasses, grasshoppers, cattle, and fences are slowly giving way to a different and less constrained world (Figure 5).

Figure 5 This old meadow at the head of one of the Sedgwick Reserve's valleys is now protected from cattle. The meadow is starting to undergo successional changes from open grassland to coastal scrub. Eventually it will become open coastal oak forest. As native grasses and bushes reinvade, they increase the number of paths that energy can follow. The increasing variety of plant species provides opportunities for a greater variety of grazers and browsers, and this in turn provides a greater choice of prey for the hawks, coyotes, and owls that occupy the top of the food web.

Odum's definition of an ecosystem suggests that energy flow supports diversity, but it does not explain how this happens. Ecologists have wrestled with this question for decades. Fifty years ago, pioneering ecologist G.E. Hutchinson pointed out that the ten percent rule limits the number of trophic levels that an ecosystem can support to four or five. Above five levels there is too little energy left (one hundred thousandth of the sun's original energy) to support predators.

But many ecosystems can support thousands of species—millions of species if one counts the "invisible world" of microbes and fungi. Hutchinson concluded that this huge number of species must be maintained by interactions that subdivide the resources of each trophic level into many parts. It is these subdivisions that increase the ecological opportunities for different species [8].

We can see this process happening in the parts of the Sedgwick Reserve that are no longer grazed, such as the meadow in Figure 5. As the grassland is gradually replaced by open forest, its plants become more diverse and its animal primary consumers have a greater choice of foods.

Many species of small rodents are multiplying in the Reserve. Merriam's chipmunks, California ground squirrels, gray squirrels, Botta's pocket gophers, and at least ten species of mice, rats, lemmings, and voles all now share this growing variety of plants. Because different rodent species specialize on different species of plants, they also help to spread those plants' seeds. As these rodent species multiply in numbers, their numbers, and those of the plants on which they prey, begin to be governed by frequency-dependent processes.

Tim Paine and Harald Beck have examined the consequences of such frequency-dependent effects in Peru's Manu National Park. Their survey was aided by the fact that large and healthy populations of many different herbivorous mammals thrive in the park.

Paine and Beck enclosed small parts of the forest, ingeniously designing the enclosures in such a way that only small, medium-sized, or large mammals could get inside them. Then they put little piles of seeds from five different local tree species in the enclosures, and measured the diversities of the seedlings that survived predation by different size classes of mammals. The results are shown in Figure 6 [9].

Large mammals had no effect on the surviving seedlings' diversity, and medium-sized mammals had a small and probably insignificant effect. But small mammals had a large and highly significant impact. Small mammals selectively removed the commonest seeds, which increased the species-richness of the surviving plants. This is because the rarer types of seedlings were no longer faced with strong competition by the commoner types, and were more likely to survive in the enclosures.

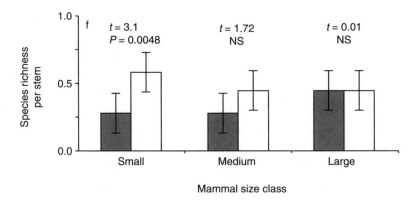

Figure 6 Frequency-dependence in herbivore–seedling interactions. When small mammals in Perú's Manu National Park are allowed to eat deliberately-placed groups of seeds in small forest enclosures, they preferentially eat the commonest seeds in each group. The dark bars show species richness in surviving seedlings when mammals of different size classes are excluded. The light bars show species richness when the mammals were permitted to eat the seeds. The result of this preferential seed-predation is that rare plant seedlings can survive disproportionately when the mammals are permitted to browse on the seeds. From part of Figure 3 of [9].

Such frequency-dependent dances of numbers, as we will see, help to maintain diversity at every level, from genes to entire webs of life [10]. And they are likely to be playing an important role in the increase in diversity of the Sedgwick ecosystem as it is released from the simplifying effects of human use.

The grazed areas of the Sedgwick Ranch are starting to revert to more complex grasslands and shrubland, and the forest is starting to spread outwards from its enclaves on the northern slopes and along the creeks that run through the property. As the shrubs and forest spread, they provide a home to the wide variety of birds and animals that have managed to survive in refuges in the nearby hills. And, as frequency-dependent interactions multiply, the Sedgwick Reserve is beginning to move towards a more diverse and more sustainable green equilibrium.

The Sedgwick Reserve is able to achieve higher diversity because the areas around it still harbor refuges where native plants and animals can persist. Tiny refuges even persist within the Reserve itself. Fifteen different native perennial grasses have managed to survive there, though all of them have been reduced to scatterings of plants that are tucked away in crannies of the foothills.

These refuges have preserved the native grasses, but they do not guarantee that the native species will survive in the face of alien invasions. Because large areas near the Reserve are still grazing land, the resourceful interloper *Avena fatua* will continue to blanket the Reserve's drier lowland hills with its seeds.

The ultimate fate of the Reserve will depend on the fate of the land that surrounds it. Currently this is a complex mix of urban, suburban, farmed, and grazed land, along with more pristine areas higher in the hills that act as refuges for wild animals and plants. It might seem impossible to forecast what will happen in such a complex and changing environment, but ecologists have managed to come up with some predictive rules.

One of these is the *Theory of Island Biogeography*, which was initially formulated to explain why islands of various sizes, lying at various distances from the mainland, tend to have different numbers of species. Scientific laws stand or fall on their predictive ability. Judged by this standard, the Theory of Island Biogeography is one of the great successes in all of ecology.

The theory had its genesis in the 1960s, when Robert MacArthur of Princeton and E.O. Wilson of Harvard began to think about how many species offshore islands could support [11]. They proposed that two things are likely to govern species numbers. The first is distance from the mainland.

They began by assuming that the mainland is the source of all the islands' species. Then they asked what would happen to two islands of the same size, one close to the mainland and one more distant. Even though the probability that a species will go extinct on both islands is the same, the closer island will accumulate more species than the distant one. This is because, over a given period of time, more mainland species can reach the closer island.

The second factor is the size of the island, which governs how quickly species are lost. If a large island and a small island are the same distance from the mainland, so that they acquire species from the mainland at the same rate, the small island should lose species at a faster rate. All other things being equal, a larger island should be able to retain species for longer.

When the rates of immigration and extinction on this second pair of islands reach a balance, the large island should have more species than the small one. Wilson and Daniel Simberloff tested this prediction by counting the number of insect species living on real islands, in this case tiny mangrove islands off the southern coast of Florida. Then they tented and fumigated the islands, killing all their resident insects.[2]

They found that after nine months the numbers of species on the islands had recovered to the numbers that were present before the islands were fumigated. Larger inshore islands, which had been home to many insect species before they were fumigated, were able to regain many species, while smaller and more distant ones that had started with fewer species were only able to regain a few [12].

Since these pioneering experiments, the predictions of the Theory of Island Biogeography have been tested repeatedly. The theory can be used to illuminate many other situations. For example, it tells us that if we want to retain many species in a park that is surrounded by human-impacted areas, we should make the park as large as possible so that its rate of species loss will be low.

Alas, this is not usually possible. The small size of the Sedgwick Reserve increases the likelihood that it will lose species over time. If we are only able to save small bits of habitat like those at Sedgwick, the Theory of Island Biogeography suggests that we should keep those small areas as close to each other as possible and establish corridors between them. Then, when Sedgwick loses a species, it may be able to regain it from a nearby area.

But such a system of reserves may not be possible. If subdivision and development of the surrounding hills continues, the long-term prognosis for the Reserve is bleak. Local extinctions will gradually deplete the native plants and animals within the Reserve. Dogs will chase the native animals. Domestic cats, which have already had a devastating effect on many of California's ecosystems, will venture into the Reserve in increasing numbers. Cats are not governed by the ten-percent rule of ecosystem energy flow, because they can go home after a hard day of hunting small animals and birds and dine sumptuously on feasts of cat food that are provided by their solicitous owners [13].

The Reserve will remain an island of ecological diversity, but the iron laws of island biogeography will determine how much diversity it can retain. And the Reserve will never return to the mix of species that was present before the Europeans arrived, much less the dazzling diversity of animals that innocently greeted the first hunters who came south from Siberia. In ecology, as in human affairs, you can't go home again.

The story of the Reserve illustrates some of the ecological laws that maintain species diversity. But where did this remarkable diversity come from in the first place? To find vivid examples of the evolutionary origin of diversity, I traveled to a more remote and far less damaged ecosystem, the rainforest of Guyana on the northern coast of South America (Plate 3).

Plate 3 shows a black caiman, *Melanosuchus niger*, launching itself into the Rupununi River through a cloud of green and gold sulfur butterflies. The Rupununi winds its way through the extensive rainforest, as yet largely undisturbed, that covers much of Guyana.

Natural Selection—Nature's *Art of War*

When my wife Liz and I arrived at the tiny domestic airport outside George-town, Guyana's raffish capital city, we were primed to do battle with draconian baggage weight limits. We began this skirmish with the airport personnel by wearing photographers' vests and filling them to capacity with our heavy photographic gear, so that we bulged in strange and unexpected places.

The next challenge was to appear utterly nonchalant as we were weighed at the check-in counter. This was old stuff for us. We had faced similar overweight luggage problems while flying to remote areas in Malaysia and Indonesia. The trick was to stand upright, just as we would do if we were not being dragged down by giant telephoto lenses. It was also important not to stagger when getting on and off the scales.

I hasten to add that we were not displacing other passengers by this subterfuge. We are both pretty skinny, and even after adding all our gear we still weighed less than the average passenger.

Our destination was the interior of Guyana, a vast region that is both extremely ancient and almost pristine. Guyana's rainforests and savannahs would provide us with startling vignettes of evolution and ecological processes in action. As we tunneled deeper into the complexity of these ecosystems and the role that natural selection plays in them, we gained a new understanding of the ways in which balances of natural selection and ecological pressures—green equilibria—maintain this complexity.

Most importantly, we discovered that if these ecosystems are to be saved, it is essential to harness the accumulated knowledge of the people who make them their home. These are the people who have adapted to their world, who depend on it, and who have the most to lose if it is destroyed.

As we began our explorations of these ecosystems we found ourselves entering an ancient conflict zone. Twenty-four hundred years ago, Sun Tzu wrote in *The Art of War* that all warfare depends on deception. He was of course talking about human conflicts, and he made the assumption that both sides in such conflicts possess at least some degree of intelligence and cunning.

In the Guyanan rainforest, the consequences of a natural version of the *Art of War* are everywhere. In nature's version the combatants must be able to deceive, but they are not always required to be intelligent. The power of natural selection can trump intelligence and cunning. This is lucky for species which, through no fault of their own, happen not to be very bright. Through the power of natural selection, even an organism that is not intelligent at all can turn the intelligence of its opponents to its own advantage.

The Iwokrama Forest covers an area of almost 4,000 square kilometers in the center of Guyana. A research center was established in 1996 at a beautiful riverside site in the heart of the Forest, under the patronage of the U.K.'s Prince Charles. At the time of our visit the center was hosting more than fifty scientists from Guyana and abroad, who were collaborating in studies of the Forest's climate, hydrology, and biology.

At a site two kilometers from the Center's headquarters, entomologist Chandra Sambhu and his colleagues are raising caterpillars of butterflies and moths that are common in the Forest, in order to monitor their diversity and explore the feasibility of butterfly farming. Soon after Liz and I arrived at Iwokrama, Chandra showed us how some of these caterpillars have co-opted the most basic instincts of their predators in order to defend themselves.

The head regions of caterpillars of the swallowtail butterfly *Heraclides thoas* carry patterns that make them look rather like a snake's head. When my shadow fell on one of these caterpillars, it responded by immediately inflating this snake-like part of its body. Although the effect was to make the snake seem to open its eyes wider, I was unimpressed. It seemed to me, as an impartial critic, that the caterpillar's efforts to impersonate a snake were falling a bit short.

But when the caterpillar was threatened more directly, it revealed a far more convincing talent for snake-mimicry. Following Chandra's instructions, I poked it gently with a twig. It instantly unfurled a red bifurcated tube that looked exactly like a snake's tongue (Figure 7). I actually recoiled in shock. It seemed that I was still in thrall to an ophidiophobia that dated back to my arboreal primate ancestors.

How, when, and why did this elaborate mimicry arise? Clearly the mimicry was driven by natural selection, and equally clearly it is too complex to have appeared in a single step.

An answer to the *why* question might seem obvious enough—snake-mimicry must surely protect the caterpillar from predation. But the hypothesis that snake-mimicry is protective has not been tested directly in the field. Other kinds of mimicry, such as the spots on butterfly and moth wings that sometimes look unnervingly like fierce predator eyes, have been shown to be extremely effective in scaring predators. But experiments have shown that other types of conspicuous patterns make the predators recoil as well, suggesting that the spots may simply provide a "startle factor" [14].

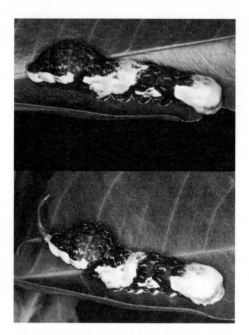

Figure 7 This caterpillar of the swallowtail butterfly *Heraclides thoas* has a swollen front end that vaguely resembles a snake's head. If this resemblance proves insufficient to discourage predators, the caterpillar has a Plan B. When it is touched, it inflates a red structure that looks like a snake's forked tongue.

I suspect that there is more to the caterpillars' snake mimicry than a mere startle factor. The caterpillars of some hawkmoths show even closer resemblances to a snake's head—but it is their tails, rather than their heads, that look scary (Figure 8). The hawkmoth mimicry must therefore have evolved independently of the swallowtail mimicry, and there must have been equally strong selective pressures governing its evolution.

The snake-head mimicries in the swallowtails and the hawkmoths represent a classic case of *convergent evolution*, in which similar structures or behaviors have arisen independently in different groups of organisms. The fact that such protective mechanisms have evolved more than once shows the benefit that caterpillars can gain by scaring the living daylights out of their predators. And the hawkmoth snake heads and swallowtail snake tongues are so realistic that I have a hard time believing that they simply represent a startle factor.

The *how* and *when* questions about these mimicries are evolutionary questions. As we trace the origin of such mimicries, we will discover how frequency-dependence plays a role in evolution as well as in ecology.

We cannot follow the evolution of caterpillar snake-mimicry in the fossil record, because fragile and soft-bodied caterpillars (and insects in general) are woefully underrepresented there. And of course the chance that a caterpillar

Figure 8 This astoundingly realistic simulacrum of a snake head, so detailed that it actually includes imitation eye reflections, adorns the tail of a hawk-moth caterpillar (*Hemeroplanes* sp.) from Costa Rica. Even caterpillars that lack imitation tongues have utterly convincing ways of protecting themselves.

might have been fossilized at the very moment that it was imitating a snake is essentially zero. But biologists have known since before Darwin's time that if one species shows an extreme structure or behavior, its close relatives will tend to exhibit less-elaborate versions of the same thing. These less-extreme versions often provide hints about how the character evolved.

Caterpillars of most swallowtail species have paired structures near their heads called *osmeteria*. These osmeteria secrete a nasty volatile compound that repels predators. In most cases the osmeteria are permanently positioned behind the head and at right angles to the axis of the caterpillar's body, so that they don't look especially like the tongue of a snake.

Because the ancestors of *H. thoas* already possessed these paired osmeteria, it must have taken a relatively small number of evolutionary steps to turn them into a structure that looked like a snake's tongue. Natural selection seized on genetic variation that was present in the caterpillar populations and that affected the size, position, shape, and color of the osmeteria. As particular forms of alleles at these genes were selected for, over many generations the osmeteria moved towards the front of the head and accumulated small modifications that made them look more like a snake's tongue. At the same time, and through a similar evolutionary process, the caterpillars' heads were selected to look more like the heads of snakes. These various selective processes resulted in increased fitness, because the most convincing of these caterpillars in each generation were the ones that were most likely to terrify predators. And the numbers of predators that could be frightened drove the selection—if they were rare the allele frequencies in the caterpillar gene pool would have changed slowly, but if they were common the rate at which the mimicry evolved would have increased.

The hawkmoth caterpillars, which lack osmeteria, have not evolved imitation snake tongues. Instead their imitation snake heads themselves have become wonderfully convincing, sometimes far more realistic than the snake heads of the swallowtails. Even though the hawkmoth caterpillars do not have that extra *coup de fouet* of a tongue that sticks out and startles, they are able to do a remarkable job in the predator-scaring department.

In order for these ruses to work, the caterpillars' bird and mammal predators must be fearful of snakes, which of course depends on the number of snakes in their environment. Given enough snakes, this phobia is a benefit to the animals that eat the caterpillars, which would otherwise themselves fall prey to snakes. But even if caterpillar-eaters become cleverer, so that they can distinguish between the real peril and a bunch of delicious caterpillars that are merely parading their thespian talents, natural selection will continue to provide an escape route for their prey. The caterpillars will be selected for more convincing ways of imitating snakes. In this natural version of the *Art of War*, Darwinian selection puts every participant, intelligent or not, on an equal footing. And it depends on the relative frequencies of all the members of the ecosystem, which in turn will drive how rapidly the frequencies of alleles change in the caterpillars' gene pools.

Darwinian evolution is of course responsible for far more than the evolution of defense mechanisms. Sometimes, paradoxically, it can select for interactions between species that seem to be the mirror image of the competition on which all evolution ultimately depends. And this type of selection can also increase ecosystem diversity.

Golden Frogs and the Evolution of Cooperation

On one of our first trips to Guyana's interior, we flew to an archipelago in the clouds, a cluster of islands made up of flat-topped mountains called *tepuis* (Figure 9). The *tepuis* have been called the continental equivalent of the Galápagos Islands. Their flat tops, isolated from each other by steep cliffs and by stretches of forest-covered plains far below, form a kind of archipelago that harbors many unique ecosystems [15].

From the top of the Potaro Plateau, one of the largest of the *tepuis*, the Potaro River hurls itself over the edge of a cliff five times as high as Niagara to form the magnificent Kaieteur Falls.

Plate 4 shows Kaieteur Falls. In the foreground is a giant tank bromeliad, *Brocchinia micrantha*.

Figure 9 Viewed from the air, Guyana's vast rainforest stretches southwards towards a distant *tepui* tableland.

Legend says that Kaie, a chief of the Patamonas tribe, paddled his canoe over the falls to his death. He hoped that his sacrifice would appease the gods and encourage his tribe to fight the fierce Caribs.

As we approached the falls we walked through a mini-forest of giant ground-dwelling tank bromeliads. These plants, which are common on the tops of the *tepuis*, are relatives of the many small epiphytic bromeliads, including Spanish moss that make their homes high up in tropical and subtropical trees. They are also related to ground-dwelling pineapples.

The bromeliads near the falls are bathed in mist, which along with the plentiful rain fills the bases of their leaves with little pools of water. These pools provide a unique home for little golden poison-dart tree frogs, *Colostethus beebei*. The adult frogs are able to hunt insects nearby, but their tadpoles must feed on the tiny organisms that multiply in the trapped water inside the bromeliads (Figure 10).

These golden frogs were first described in 1921 by the naturalist (and pioneering deep-sea aquanaut) William Beebe [16]. They have only been found in the bromeliads that live near Kaieteur Falls and that harbor permanent pools of water.

19

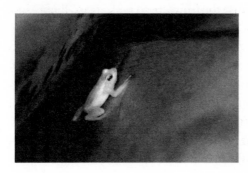

Figure 10 A little golden poison-dart frog, *Colostethus beebei*, peers out of the pool of water that has accumulated at the base of the leaves of a tank bromeliad.

The frogs and the bromeliads that live on the top of this remote *tepui* occupy their own highly specialized *ecological niches*. Their ecosystem also supports many other niches. But what exactly is an ecological niche? Although we tend to bandy the term around freely, and we all have at least a vague idea of what it means, it is worth taking a close look at this central ecological concept.

The word niche in its ecological sense first appears in a 1917 paper by ornithologist Joseph Grinnell. He was studying the habits of the California thrasher *Toxostoma redivivum*, a species of shy insect-eating songbird that lives in the Pacific Southwest [17]. Grinnell used the term niche in the sense of the French verb *nicher*, meaning to nest or to dwell. A niche, he proposed, is the physical and biological environment of a species.

A decade later, English ecologist Charles Elton adopted Grinnell's term but took a different approach to defining it. He emphasized the ways in which organisms use their immediate environment. An organism's niche, he proposed, consists of its relations to its sources of food and to its enemies [18]. Two different species might carve out different ecological niches from the same set of environmental factors.

It is clear that both of these definitions are partly right. Any useful definition of a niche must take into account both the capabilities of a species and the environmental factors that it must face. A blending of these two definitions explains why some ecological niches are far narrower and far more highly specialized than others. Some species, like red-winged blackbirds that can eat both seeds and insects, are able to utilize a wide range of the resources that are provided by their environment. More specialized species have evolved to utilize a narrower spectrum of the environment's resources.

The golden frogs inhabit an especially narrow Grinnell-type niche, made up of the tiny part of the *tepui's* physical environment that they have been able to exploit in a highly specific way. Because the frogs are so rare, and so highly specialized, they are also at great risk of extinction. The wet area around the waterfall, a few hectares in size, supports only a few hundred of the giant bromeliads that can provide homes to the frogs. The rules of island biogeography state that if a species is confined to such a small area it has a high probability of going extinct. Something as seemingly minor as a spell of unusually dry and hot weather that lasts for a year or two might wipe out all of these frogs.

In spite of this grim prognosis, the power of natural selection allows the frogs to rise to the challenge. As Elton realized, it is this interaction between adaptation and ecology that has allowed species to shape their response to the physical and biological characteristics of their environment. Evolution can give even the rarest species, like the frogs, a fighting chance. The rarest species in an ecosystem have survived in part because they have accumulated abilities that allow them to alleviate the disadvantage of rarity. Those that have not survived, alas, were not as successful in this evolutionary race.

The *C. beebei* frogs have adapted in many ways to build and modify their narrow ecological niche. One of these ways is to alter the mechanisms by which their bodies respond to environmental change.

Droughts pose the greatest danger to the frogs. When water levels fall in their little bromeliad reservoirs, as often happens even during a normal dry season, there is not enough food for their tadpoles. During these dry periods the female frogs encourage a kind of benign cannibalism, laying unfertilized eggs on which their already-hatched tadpoles can feed [19].

This reproductive strategy works because it builds on the special *symbiotic* ("living together") relationship that has already evolved between the frogs and the bromeliads. Such symbiotic interactions among species provide a rich source of ecological niches, often drawing even more species into the symbiotic relationship.

There are three different kinds of symbiosis: *commensalism*, *mutualism*, and *parasitism*. Commensalism means "living at the same table," in effect sharing the same resources. In a commensal symbosis, one species provides an ecological niche for the other, without being harmed in the process. It might at first appear that the interaction between the frogs and the bromeliads is commensal, because one species (in this case the frogs) benefits and the other species (the bromeliads) is unaffected. Many of the bromeliads on the tops of the *tepuis* do not harbor frogs. Thus, the bromeliads can survive nicely in the absence of frogs but the reverse is not true.

Actually, it is almost impossible to demonstrate that a relationship is truly commensal, because it is difficult to prove that one of the species is truly unaffected by the activities of the other. And indeed it has now been shown that the frog–bromeliad symbiosis is a two-way street. Feces deposited by the frogs in the trapped pools of water fertilize the bromeliads, increasing their rate of photosynthesis [20].

The frog–bromeliad symbiosis turns out to be, not a commensalism, but rather a mutualistic symbiosis in which both species benefit. If these appealing little frogs were to go extinct, the bromeliads would survive. But if the small number of bromeliads that grow near the falls were to be wiped out, the frogs would disappear. Although the mutualism is thus far more beneficial to the frogs than it is the bromeliads, the nourishment provided by the frogs gives the bromeliads an extra margin of safety in their own struggle for survival. But we will soon encounter other mutualisms in which the participating species absolutely depend on each other for survival.

Darwinian evolution ultimately depends on competition between individuals and between species, in an unending battle for limited resources. Symbiotic interactions, like those between the frogs and the bromeliads, can only evolve if they aid the collective competitive abilities of the participants in the symbiosis. Taken together, the team of frogs and bromeliads is better able to face the world's perils than either species could have done alone.

Landscape With Water Lilies

Competition is the ultimate driver of evolutionary change. But mutualistic interactions, such as those between the frogs and the bromeliads, increase their ability to cooperate, not to compete. This puzzle resolves itself when we consider the unforgiving world that lies beyond the species that cooperate mutualistically. If cooperative interactions help to make species better competitors against the rest of the world at large, then species characters that lead to teamwork will be selected for.

The world's most magnificent water lily, *Victoria amazonica*, plays a role in many such mutualistic interactions.

In search of some of these water lilies we left Yupukari Village at the edge of Guyana's rainforest and traveled by dugout along the Rupununi River. From the river we followed a narrow tributary, sheltered by overarching trees, to a nearby oxbow lake that was almost completely covered by lily pads.

The story of the lilies begins with the Rupununi itself. Along with the many other rivers that wind like snakes through Guyana's tropical low-lands, the Rupununi supplies the water lilies with new oxbow lakes in which they can survive. The rivers carve away the land at the outer banks of each of their curves, where the water flows most rapidly. Over time this makes the curves more pronounced, so that the rivers become more and more sinuous.

Eventually adjacent loops of a river almost touch, separated by only a narrow neck of land. When the flow breaks through this narrow point it shortens the river's path, leaving behind a C-shaped stretch of water with no outlet. This orphaned bit of river has suddenly become an oxbow lake, a kind of fossil that preserves a fragment of a route that the river once followed.

We drifted on the oxbow lake's calm surface under a cloudless sky. The lilies were in the midst of their flowering season. On the first evening of the flowers' short lives they shine brilliant white.

In Plate 5, a newly-opened lily flower floats like a reflection of the moon beneath the evening sky. A Tupi-Guarani legend tells of a princess who fell in love with the moon and chased it vainly, until the moon deigned to notice her and turned her into a water lily. Whenever she flowered, she became a reflection of the god she loved [21].

Giant water lilies are found throughout the Amazon basin, but the Guyanan species is by far the largest. Its leaves can reach more than two meters across. One of the most famous photographs of these lilies features a lustily crying baby supported by one of the pads.

The lilies play a role in many commensal and mutualistic interactions. At the commensal end of the spectrum, the lilies' roots and decaying leaves form a rich tangle that provides shelter for the young of many fish. And a number of bird species nest on the pads, or use them as floating platforms from which to hunt fish (Figure 11).

As the evening grew darker, sharp zinging noises heralded the arrival of the lilies' most important mutualistic symbiotes, *Cyclocephala hardyi* beetles. The little brown beetles landed on the open flowers and dove into their interiors (Figure 12). There they were welcomed by a gentle warmth, generated by the sped-up metabolism of the flowers as they matured and opened [22].

As darkness fell the flowers closed, trapping the beetles. The flowers' heat encouraged the beetles to stay active during the night, so that they could cover themselves thoroughly with pollen. It is during the night that the beetles benefit

Figure 11 A wattled jacana, *Jacana jacana*, strides across the carpet of plants that cover the oxbow lake. She has laid her eggs on this water lily pad, and is watching them assiduously. The lily pads provide nesting space and hunting platforms for a dozen species of bird without being harmed themselves, a classic example of a commensal symbiosis.

from the mutualism. The lily flowers are so large and complex, with so many carpels and anthers, that they can stand some loss when the beetles dine on their tenderer parts.

When the flowers open in the relative cool of the following morning they will release their little captives. The flowers will even bestow a going-away present—an extra burst of heat as the beetles depart with their loads of pollen.

The oxbow lake that we drifted across that evening will vanish in a few decades, along with its population of lilies. The lake will fill up with a mass of lily roots that will eventually become soil. But as long as Guyana's rivers are able to wriggle unimpeded across the flat lowlands they will spawn new oxbow lakes. Lily seeds and bits of root will soon spread to these new lakes, producing new lily populations. These in turn will re-establish the many commensal and mutualistic symbioses that add ecological niches to this enchanted landscape, and that help to maintain its diversity.

Figure 12 A pollen-covered *Cyclocephala hardyi* beetle, visible towards the upper left, explores the base of a water lily's petals before it disappears into the cozy, anther-lined hole in the center of the flower.

Commensalism and mutualism are essential components of the green equilibrium of this ecosystem. But as Liz and I probed deeper into the forest's ecological complexity, we began to uncover another component of this equilibrium. This was the role that is played by the invisible world.

Pathogens, Parasites, and the Invisible World

A single hectare of the most diverse tropical forests, such as Lambir Hills in the Malaysian province of Sarawak or the Yasuni Forest of Ecuador, may be home to almost a thousand different species of tree and vine. This dizzying plant diversity supports hundreds of species of birds and mammals. Ecologists have battled over the reasons for such high tropical diversity. Likely explanations include the huge amounts of energy from the tropical sun and perhaps unusually high rates of evolution in the tropics [23]. There is, however, uncertainty about the latter: tropical extinction rates seem not to be unusually high [24].

Much evidence is now emerging that mutualism and commensalism add many ecological niches to ecosystems. The benefits of living together allow two or more species to coexist where there might have been an ecological niche for only one species without the symbioses. Such symbiotic interactions help to explain why there are so many species, especially in the tropics where seasonal change is slight and the living world can bloom and multiply all year round.

There is another huge source of ecological niches. All ecosystems support an invisible world of pathogenic, parasitic, and mutualistic microorganisms that have carved out their own niches. Although parasites and their hosts live together symbiotically, they are different from mutualistic species which confer benefits on each other. Parasites rob their hosts of resources and bestow only harm in return.

In 1962, Jan B. Gillett, a botanist working in Africa, suggested that pathogens and parasites make ecosystems more complex by their very presence. Because of parasitism, he pointed out, there are far more invisible species in an ecosystem than visible ones. Each of these invisible species occupies an ecological niche—at the expense, of course, of its host [25].

This invisible diversity also provides a major explanation for why most species in the visible world are rare. In the most diverse ecosystems, such as tropical forests, no single species of animal or plant predominates. Pathogens and parasites help to explain why.

Rupicola rupicola, the Guiana Shield's unique species of Cock-of-the-Rock, is an iconic example of rarity in the visible world. These birds are so shy and rare that I had to make five attempts before I could get satisfactory photographs of them.

Cocks-of-the-Rock live near (and get their name from) isolated groupings of granite slabs that are scattered through Guyana's forest like stonehenges that have lost their way. The slabs are ancient intrusions into softer rock that has eroded away, leaving the harder granite behind. Their undersides now provide safe places for the birds to build their nests.

The male Cocks-of-the-Rock carefully sweep little clearings on the forest floor to make *leks* or display sites, where they dance and show off their plumage to highly critical audiences of females. At the same time they provide a fine example of Charles Elton's definition of an ecological niche. They defecate many different kinds of seeds during their displays. The leks soon become surrounded by trees and bushes that provide a rich source of food and that are much more diverse than most of the forest. These food plants may be at least as much of an attraction for the females as the grandstanding of the males.

I visited two lek sites, one of them three times, only to find them empty of birds. Finally, after a nine-kilometer hike from the village of Wowetta through dense forest, my guides and I arrived at yet another of the hidden clusters of rock slabs.

There were no birds at the likely lekking areas near the rock slabs, but we did spot several of the brilliant orange-red males, along with a couple of the shy brown females, flitting among the trees in the open forest. They were too far away to photograph. In the hope that they might come closer if we hid from view, we crept into a dark opening between leaning slabs of rock.

We moved cautiously in the gloom, climbing as quietly as possible over the uneven stone slabs that made up the floor of the cave. Our way was lit by small openings between the rocks that gave glimpses of the sunlit forest beyond. Suddenly a male Cock-of-the-Rock appeared and posed elegantly in one of these openings, framed by the flat slabs of ancient stone that form its home and that provide its eponymous name (Plate 6). The bird's orange-red plumage, distinctly patterned wing primaries, and endearing tutu of frilly covert feathers distinguish it clearly from the redder Andean species R. *peruvianus*.

There are many reasons why these birds are rare, but one of the most fascinating is a frequency-dependent one. Rarity is sometimes protective. It may protect the birds from the parasites that prey on them.

Parasites are everywhere. When poachers kidnap parrots and macaws from the South American tropics and smuggle them to the pet shops of the developed world, the birds are not alone. They are accompanied by lice, fleas, ticks, mites, and bugs, and by a wide variety of bird malarias and other blood parasites (Figure 13). They can harbor the bacterium that causes psittacosis or parrot fever, the fungus that causes the lung disease aspergillosis, and a variety of dangerous viruses. Some of these avian diseases have made the jump to humans.

In a world free of such parasitism, there would be no horde of tiny creatures to bite, burrow, chew on, suck at, and lay their eggs in the shrinking flesh of the ecosystem's larger inhabitants. And there would be none of the less obvious organisms that swarm in animals' bloodstreams and that infect the leaves and rot the roots of plants.

All these teeming parasites are invisible to us when we encounter their host animals and plants in the wild. Yet they have a great influence on where their hosts can live and on their hosts' ability to survive. Their influence depends in part on how rare or how common their hosts are.

If a species is rare, like the Cock-of-the-Rock or the golden frog, its very rarity puts it at risk. Rare species that occupy narrow ecological niches are less likely to survive than more numerous species that occupy wider ecological niches. But

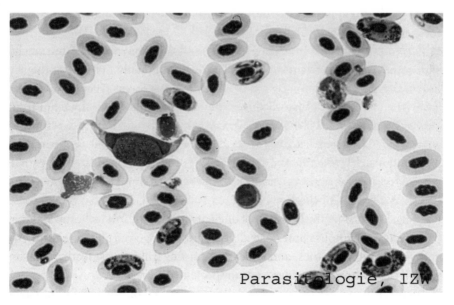

Figure 13 Some of the thousands of different protozoan parasites that have been found in the blood of wild bird populations. In this blood smear from an owl, the small parasites inside the red cells are *Haemoproteus*, and the huge free-living parasite is a *Leucocytozoan*. Most of these infections do not cause obvious symptoms, but they will inevitably take a toll on stressed or malnourished birds.

rarity may also confer a benefit, because rare animals and plants are less likely to pick up diseases and parasites from other members of the same species.

Like other frequency-dependent relationships, this frequency-dependence of host–parasite relationships helps ecosystems to remain stable. Host species that are rare are likely to have a small parasite load, allowing them to thrive in spite of their rarity. But when host species become common, their parasites spread more easily. The multiplying parasites help prevent their host species from becoming so numerous that they drive other species to extinction.

This pattern of selection, known as *negative frequency-dependence*, is distinct from *positive* frequency-dependence, in which a species becomes fitter as it becomes more common. A species that is positively frequency-dependent, like the invasive *Avena* grasses of California, can modify its ecological niche as it expands in numbers. These grasses drive out other similar species such as the native grasses instead of coming into balance with them.

Even when they are rare, tropical birds such as the Cock-of-the-Rock are not disease-free. Their blood is likely to contain several kinds of parasite. In addition to avian malaria, parasites of six other great groups of single-celled pathogens have been found in birds. They include the bird equivalent of sleeping sickness, and a voracious eater of red blood cells called *Haemoproteus*. As if this were not enough, many species including the Cock-of-the-Rock harbor tiny filariid worms that are the avian equivalent of the worms that cause human elephantiasis [26, 27].

In spite of this zoo of parasites, few birds in the wild probably die of their effects. The cock that I photographed was healthy and alert, in part because it had coevolved with its parasites so as to reduce their impact. Parasites of rare hosts put themselves in especial danger if they make their hosts too sick, because the chance that their descendants will be passed to a new member of the host species is low. Nonetheless, even relatively benign parasites can extract a toll. When a bird is malnourished or stressed, even a small load of parasites will reduce its chances of survival.

In this chapter we have encountered intense competition among plants for sunlight and water, competition among predators for prey, strong selection for the ability of prey organisms to hide from and repel predators, symbiotic interactions that add to the capabilities of many species, and the complex world of pathogens and parasites that helps to drive ecosystem diversity. All these mechanisms, and many more, help to maintain the green equilibria of ecosystems. And all of them have originated through the process of Darwinian evolution. Natural selection continues to adapt every species to its incredibly complex environment, drawing on the genetic resources of the species' gene pools and sometimes generating new species in the process. Every ecosystem is a work in progress.

Genetic Equilibrium

The golden frogs have adapted in ways that help them to survive in spite of their rarity. But such evolutionary adaptation takes both time and genetic resources. Many of these genetic resources have accumulated because of frequency-dependence. These equilibria are just as essential to our understanding of the living world as the ecological equilibria that we have already encountered.

The gene pools of almost all species are filled with genetic variation. As we saw in the introduction, each of a species' thousands of genes may be represented in its gene pool by several different allelic forms that have been generated by mutations. Genes that are present in populations in more than one allelic version are called *polymorphic* ("many forms").

Some individuals in a population are *homozygous* for one allele of a gene and others are homozygous for another. *Heterozygous* individuals carry two different alleles, one inherited from their mother and one from their father. But, as Figure 14 shows, no individual can carry all of the population's different alleles.

Both golden frogs and humans have about 25,000 different genes, and the gene pools of both species carry more than one allelic form of most of these genes. Many other parts of the genome are also important in determining how and when the various genes are expressed, and these regions also have alleles. Each generation, the alleles are recombined in myriad ways by the mixing of genes from the two parents that takes place during sexual reproduction.

Such copious amounts of genetic variation open up almost infinite possibilities. To get some notion of this, we can calculate the probability that two unrelated individuals chosen from the same genetically diverse population will have the same complement of alleles, the same *genotype*.

The probability of such a match is truly infinitesimal. The genes in the human (or the frog) gene pool carry alleles at a wide variety of frequencies. Let us begin our calculations by making the not unreasonable assumption that there are a hundred genes in the human gene pool that each happen to have two alleles at frequencies of 0.7 and 0.3.

At just one of these genes, the chance that two unrelated individuals will have the same *genotype* is 0.42. This is the sum of three probabilities. The first is that

Generation of Diversity
Because of Polymorphism
and Genetic Recombination

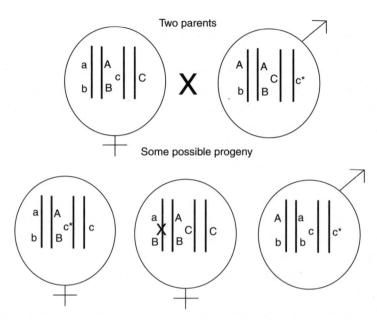

Figure 14 This figure, which shows two mated individuals and three of their progeny, demonstrates how genetic polymorphism and genetic recombination can generate variation in a gene pool. The three genes on two chromosomes that are shown in the diagram represent a tiny subsample of the 25,000 genes and 23 chromosomes in each human. In the parents at the top of the figure two of the genes, A and B, are represented by two alleles, and the third gene, C, is represented by three alleles, C, c, and c*. Because the parental chromosomes assort randomly to the progeny, each of the progeny carries a genotype different from those of the other progeny and from those of their parents. In addition, the second child carries a chromosome that arose by recombination in her female parent (marked with an X). In a polymorphic population, random assortment of chromosomes and genetic recombination work together to produce a vast array of new genotypes each generation.

they will both be *homozygous* for the first allele—that is, that they have two copies of the allele. This probability is 0.7 × 0.7 × 0.7 × 0.7 = 0.24. The second probability is that they will both be *heterozygous* for the two alleles—that is, that they will carry one copy of one allele and one copy of the other. This probability is given by the chance that the first individual will be heterozygous, which is 2 × 0.7 × 0.3, multiplied by the chance that the second will be heterozygous, which is also 2 × 0.7 × 0.3. This second probability works out to 0.176. The third probability is that both individuals will be homozygous for the second allele, which is $0.3 \times 0.3 \times 0.3 \times 0.3 = 0.008$. The three probabilities sum to 0.42. The other 58 percent of the time, unrelated individuals will have different genotypes for this gene.

Thus, there is a close to fifty percent chance that the two individuals will have identical genotypes for this particular gene. But remember that we are dealing with a hundred genes, which are inherited largely independently of each other so that they can come together in many different possible combinations. The chance that two people will have the same genotype at two of these genes is 0.42 × 0.42 or 0.176. For ten genes, the chance that all ten will carry the same genotypes in both individuals diminishes to 0.42 raised to the tenth power, or 0.00017. The chance of identity at all hundred genes drops with breathtaking rapidity to 0.42 raised to the hundredth power, or 2×10^{-38}.

Even though we have confined our attention to this small subsample of our genes, we already have a probability of identity between individuals that is infinitesimally small. No two people (except for identical twins) would ever have carried the same genotype at these hundred polymorphic genes.

Estimates vary, but it has been calculated that a total of about 100 billion or 10^{11} modern humans have lived on Earth since the origin of our species [28]. We would need almost 10^{29} times as many people, a total of a hundred thousand trillion trillion times 100 billion, to reach the number 10^{38}. Only then would we have a good chance of finding two people with identical genotypes at these hundred polymorphic genes.

If we extend such calculations to the rest of our 25,000 genes, and to the many alleles that are found in other parts of our DNA, the probability of genetic identity between two individuals (once more excepting identical twins) drops essentially to zero.

Recall that these various alleles have arisen by mutations. Mutations are changes that take place in our DNA and that are then inherited through subsequent generations. Such changes can alter genes, and can alter how genes are expressed, in many different ways. Most of these mutant alleles are harmful or have little effect, and most such alleles soon disappear from our gene pool. But the small fraction of mutant alleles that do survive provide an enormous resource for evolutionary change. The resource becomes especially valuable if the environment alters in some unpredictable way.

As one generation succeeds the next, the best-adapted individuals outreproduce the rest. When the environment shifts, some alleles that formerly were beneficial or neutral in their effects suddenly become harmful. Other alleles may switch from being harmful or neutral to conferring an advantage to their carriers. Such newly-favored alleles, once rare, will be pushed up in frequency by the process of natural selection, while the newly harmful alleles will become less common. As favored alleles of different genes rise in frequency, they are able to come together in brand-new combinations, some of which will result in even better-adapted members of the population.

As we will explore in the next chapter, frequency-dependent selective pressures play a major role in the adaptation of populations of frogs, humans, and every other species. Potentially useful genetic variation can accumulate rapidly in a gene pool if some alleles are advantageous when they are rare and lose that advantage when they become common. When such frequency-dependent selective pressures act on alleles, their ability to maintain genetic variation is remarkably reminiscent of the frequency-dependent interactions that maintain the balance among the many species of predators and prey in an ecosystem like the Sedgwick Reserve.

But any potential advantage that might be provided by the genetic variation in a gene pool is nullified if the right variation is simply not there. Because the gene pool of the golden frog's tiny population is so small, it might not possess a critical new allele that could be essential for adaptation when its environment changes. The fate of this frog species would then become hostage to the fickle and unpredictable process of mutation. The frogs might go extinct before the needed mutation appears, especially if the environmental change is rapid.

Such an environmental peril for the frogs may indeed be looming, not because of immediate human impact, but because the surrounding forest is slowly expanding and trees are starting to grow closer to the falls. As the trees' leaves shower down onto the bromeliads they block the frogs' access to their water reservoirs. Then, as the leaves decay, they contaminate the trapped water and kill the tadpoles' food supply [29].

Are these trees able to invade because human-caused climate change is making the tops of the *tepuis* warmer and wetter? The jury is still out on this possibility, but even if we cannot be blamed for this particular change in the frogs' environment there would still be many reasons why these frogs, like every other rare species, are continually confronted with the threat of extinction.

The evolutionary predicament of these frogs is strikingly different from the evolutionary flexibility of a very different population that we encountered in the first part of this chapter. When I hiked over the golden hills of the Sedgwick Reserve, I was struck by the remarkable color match between the introduced

Avena grasses and the native grasshopper *Melanoplus devastator*. I wondered whether, since *Avena* began to turn the hills golden two centuries ago, the grasshoppers have evolved to match the color of the *Avena* florets.

The answer is likely to be yes. The enormous populations of these grasshoppers can evolve quickly, and are able to track human-caused changes to their environment. The color and size of the adults, and the speed at which the grasshoppers pass through their many developmental stages, all vary greatly depending on the grasses on which they feed [30]. The evolution of the colors of the immature stages has not yet been investigated, but I expect it will be found to be equally variable across the grasshoppers' range and equally dependent on the local environment.

These Californian grasshoppers, unlike the golden frogs of Guyana's *tepuis*, occupy a broad ecological niche. They also possess an immense gene pool that we have made even larger because we have increased their population size by providing them with rich sources of food. The grasshoppers' huge collection of genes includes many mutant alleles that give them the potential to survive even when their environment changes rapidly in unexpected ways.

Interplays of evolutionary pressures, genetic resources, ecological complexity, and environmental change govern the fates of every rare and common species in an ecosystem. The pool of genetic variation that a species can draw on, some of which is maintained at equilibrium points by a balance of selective forces, will ultimately determine whether the species can adapt and survive. In subsequent chapters we will see how these processes have shaped our own survival as strongly as they have shaped the survival of the golden frogs and of the devastating grasshoppers.

Our tours through the Sedgwick Reserve and the Guyana rainforest have given us some insight into the laws that have shaped the evolution and the maintenance of ecosystem diversity. As we delve into these processes in more detail, we will see how they govern both the negative and the positive impacts that our species is having on the living world.

CHAPTER TWO

Maintaining a Green Equilibrium

In early 2011 I had the opportunity to dive intensively at over twenty sites scattered through the center of the Philippine archipelago. The experience was alternately wonderful and grim, illustrating both the resilience of these marine ecosystems and the points at which their resilience fails (Figure 15).

My trip began at Anilao, a small municipality about two hours south of the sprawling capital of Manila. Anilao serves as a jumping-off place for a wondrous variety of dive sites that are scattered along the coasts of the nearby Calumban Peninsula and Maricaban Island (Figure 16).

Anilao is home to many small Philippine-run dive resorts, and each weekend hundreds of divers descend on the reefs. In addition to the pressures from tourism, runoff of excess chemicals and fertilizers from the agricultural land to the north of the peninsula flows into the two great bays of Balayan and Batangas that flank the peninsula. But the coast of the peninsula itself is precipitous and still heavily wooded, so that it forms a buffer between the reefs and the agricultural area.

Frequency-Dependence Shapes the Underwater World

My first dives, with underwater photographer and nudibranch expert Mark Strickland and local guide Mark Castillo, were at tiny Sombrero Island, a site off the western tip of Maricaban. This shallow reef is carpeted with a wide variety of hard and soft corals and dominated by towering two-meter-tall sponges. Dense schools of little peach-colored fish of the Anthia subfamily hover over the reef (Plate 7).

And yet there are small but worrying signals of ecological instability even in this seemingly healthy and untouched world. The top predators on this reef are missing. Their absence reminds us of one of the central ecological rules that we explored in the last chapter, which is that the presence of a variety of top predators is essential for the maintenance of ecological diversity.

Figure 15 An orange anemonefish, *Amphiprion sandaracinos*, fans fresh seawater across its eggs on one of the lush reefs of the Anilao region of the Philippines. The eggs are just about to hatch, and the embryos' eyes are clearly visible. One might be tempted to suggest, anthropomorphically, that the babies and their parent are gazing adoringly at each other, but this is a temptation that I will resist. The reefs of Anilao are remarkably undisturbed and rich in such amazing vignettes, even though they lie just two hours away from the capital city of Manila and tourists flock to them. These reefs are lucky to possess a robust green equilibrium, in which a balance of evolutionary and ecological forces actually maintains and increases their diversity and resilience.

At the Sedgwick Reserve, the top predators include the coyotes, mountain lions, and eagles that have managed to survive from California's lost wild world. At the Sombrero Island reef, the top predators once included sharks, tuna, and barracuda.

Predators concentrate their attention on the prey species that are commonest and easiest to catch, driving them down in numbers. Once these prey have become so uncommon that they are hard to find, the predators switch their efforts to prey species that were formerly rare but that have become common in the meantime. As these prey are driven down in turn, the cycle continues.

It is difficult to measure such predator–prey interactions in detail in the natural world, but it is possible to re-create them under controlled conditions. In 2002, Alan Bond and Alan Kamil of the University of Nebraska used captive blue jays in

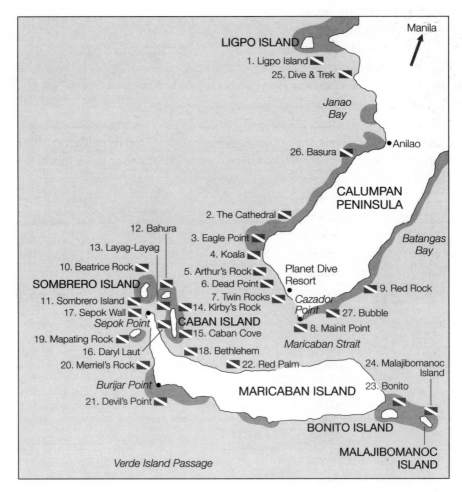

Figure 16 This area 100 kilometers south of the Philippine capital of Manila is heavily impacted by tourism, but its underwater ecosystems are still thriving.

a laboratory setting to demonstrate the effects of negative frequency-dependent selection on predation. They trained the birds to watch patterns that were flashed for a few seconds on a computer screen. The birds' task was to find images of moths in the patterns, even though the moth images were camouflaged on a busy background. When the jays found what they thought was a moth, they pecked at it. If they had guessed right, they were rewarded with a pellet of food [1].

The birds, it turned out, were unable to focus on more than a few types of search image at any one time. They quickly learned to detect the wing patterns of the types of virtual moths that they were shown most often, a skill that enabled them to get the largest number of rewards for the smallest intellectual effort.

But Bond and Kamil had designed their computer program to respond to the birds' learning curve, in the same way that prey populations would respond in the real world. The program manipulated its store of virtual moths in response to the birds' hunting patterns. During the course of the experiment the initially common types of moth, the ones that the birds were finding most easily, were flashed less and less often on the computer screen, and formerly rare types turned up more often. The birds soon switched their attention to the moths that had been rare but were now becoming commoner.

The computer also periodically generated new types of moth, mimicking the role of mutation in the real world. As the experiments proceeded, the wing patterns of the population of virtual moths that survived in the computer's memory were the ones that best matched the background pattern, making them harder for the birds to detect. The moths also became more and more diverse. Many different types of moth were able to coexist, because as each type began to rise in frequency the birds learned to recognize it and drove its numbers down again. Over time the population of virtual moths reached a diverse equilibrium of many different moth types, each regulated by negative frequency-dependent selection as the jays continually shifted their search images (Figure 17).

The jays exhibited frequency-dependent predatory behavior in the laboratory. Are such behaviors taking place in the real world? On the Great Barrier Reef's Lizard Island, predatory fish have been shown to track the abundance of their prey. Although predator–prey interactions could not be observed directly, the predators were found to move to the parts of the reef that had the largest number of prey fish [2].

The Sombrero Island reef may already be suffering from a reduction in such frequency-dependent predation. In the course of several dives we saw many different small reef fish species, but the anthias predominated. Our guide Mark Castillo had helped his father fish the reefs as a child. Now he dives on them daily and is at the forefront of the reef protection efforts that have been initiated by the local villages. He told us that ten years earlier the reefs had been home to many sharks and large pelagic fish such as tuna and barracudas. Now most of these top predators have gone, perhaps allowing the anthias and other fecund reef fish to multiply unchecked and throw the ecosystem subtly out of balance.

Possible evidence for such an imbalance has been found at another western Pacific reef near Indonesia's Bintan Island, just southeast of Singapore. From

b Experimental results (F$_{100}$)

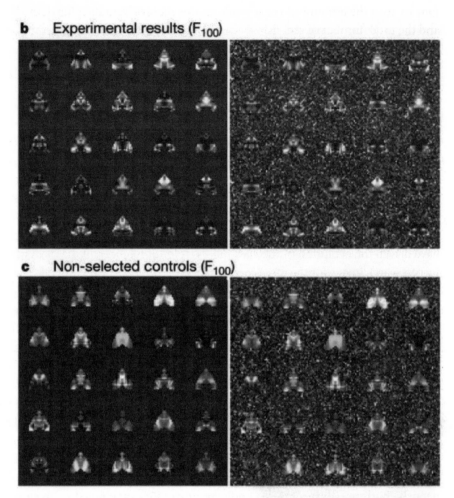

c Non-selected controls (F$_{100}$)

Figure 17 This figure, from Figure 1 of the Bond and Kamil paper, shows how populations of virtual moths have become diverse as they were hunted by trained blue jays and as the computer program changed the moth frequencies in a frequency-dependent fashion in response to the jays' shifting search images. The pictures on the left show the moths against a plain background, and those on the right show how the job of the blue jay "predators" was made more challenging when the moth images were set against a noisy background. As selection proceeded, the moths became both more diverse and more difficult to detect. In control experiments (bottom), moths were simply drawn at random from the computer's store of moths throughout the experiment, so that the jays' frequency-dependent hunting capabilities had no effect on the population of moths. These control moth populations remained less diverse and easier for the birds to find throughout the experiments.

1993 to 2010 the amount of coral cover on this reef has remained unchanged, and the total number of reef fish species has also stayed constant. But some of these species have become much commoner and others much rarer over these two decades [3].

A Brief History of Frequency-Dependence

Frequency-dependent selection has been known for more than a century. In 1884, British entomologist E.B. Poulton noticed that some Geometrid moth species have two colors of caterpillars: green and brown. The green caterpillars are well-disguised when they are on fresh young twigs, while the brown ones blend in more effectively when they are on older and browner twigs.

Was it possible, Poulton wondered, that birds that prey on these caterpillars have driven the evolution of the caterpillars' color differences? If the birds can form a search image for the most conspicuous caterpillars, they will drive them down in frequency and let those with the more cryptic color increase [4].

Remarkably, as the caterpillars grow through successive developmental stages or *instars*, they are able to switch from green to brown and back again [5]. Later studies confirmed Poulton's conclusion that the caterpillars can use visual clues from their environment to trigger these switches [6]. But once the caterpillars have committed to a color, they are stuck with it until they molt their skin and move to their next instar.

Like cuttlefish and octopuses, all of these Geometrid caterpillars carry genetic equipment that allows them to change their appearance to track changes in their environment. But unlike cuttlefish and octopuses, which can match their backgrounds within seconds, the caterpillars are unable to make the switch instantaneously. Birds are always able to find caterpillars that have failed to catch up with their changing backgrounds. As the seasons change and the world shifts from brown to green and back again, frequency-dependent hunting behavior by birds continues to push the caterpillars towards a better ability to track and match their environment.

There is a term for this—*polyphenism*. This is the ability of an organism, like Batman's arch-enemy Clayface, to alter its *phenotype* or appearance in different ways during its lifetime.

Since Poulton's pioneering observations, many other consequences of frequency-dependent predation have been found. Often these lead, not to a Clayface-like polyphenism in the prey, but to balanced genetic polymorphisms.

Females of the spectacular swallowtail butterfly *Papilio dardanus* from East Africa show many different wing shapes, colors, and patterns. They can mimic at least 25 other butterfly and moth species. The top of Plate 8 shows males and females of the swallowtail butterfly *Papilio dardanus* from the island of Madagascar, where the females do not mimic distasteful species. Below on the left are a variety of distasteful butterfly and moth species from the African mainland. On the right are the *P. dardanus* females that live in the same regions and that mimic them.

Birds find the *Papilio* butterflies delicious, but the species that *Papilio* mimics taste terrible and some of them are poisonous. The presence of these distasteful *model species* protects the tasty *Papilio* mimics—but only if predatory birds are smart enough to remember that a butterfly of a particular color or pattern tastes bad.[1]

This is a classic case of *Batesian* mimicry, named after Henry Walter Bates, the nineteenth-century naturalist who first described it in Brazilian butterflies. Examples of Batesian mimicry have now been found in most ecosystems.

The adult *Papilio* butterflies are not polyphenic. They have no opportunity to change their wing colors and patterns once they have emerged from the chrysalis. Instead, *P. dardanus* has been selected to be genetically polymorphic—its gene pool is polymorphic for alleles of genes that control wing color and pattern.

Papilio dardanus provides a dazzling example of how Batesian mimicry can contribute to genetic equilibria through frequency-dependent selection. This butterfly species is found over a large area of East Africa, and as far afield as the island of Madagascar. On that isolated island there are no distasteful butterflies that can provide good models for *P. dardanus*. This is why, as we can see from the top pictures of Plate 8, the female and male *P. dardanus* look the same [8].

In Africa the situation is very different. As many as six different nasty-tasting model species may inhabit a given part of the *P. dardanus* range. In that same region *P. dardanus* females are found that mimic all of them. But if a model species is absent, none of the *P. dardanus* females will mimic it.

If a distasteful model species is present in large numbers, the *P. dardanus* females that mimic it will tend to be common as well. This is because they are able to bask in the protection that is provided by the abundant model. Conversely, if the model species is rare, so are the mimics.

All of the *P. dardanus* butterflies in an area interbreed freely with each other, but in spite of all this genetic shuffling the complicated female mimic patterns are somehow inherited intact from one generation to the next. How can this happen? Surely there is no single gene in these butterflies that has different alleles, each of which can simultaneously affect wing color, pattern, and shape.

The answer lies in gene regulation. Distinct alleles of a single gene called H have been found in the different morphs of *Papilio* females. The H gene is a highly sophisticated regulatory gene, a so-called *master regulator* that is capable of switching on or off entire groups of genes that control different developmental pathways [9]. Each H allele switches on a different pathway in the females, and each of these pathways leads to an adult with a distinct type of color and pattern. The frequencies of the different alleles of H, and the details of the developmental pathways that are under their control, have all ultimately been shaped by predation. Clever birds, with their excellent memories, prey on the *P. dardanus* populations in a frequency-dependent fashion, and the result is a genetic polymorphism of astonishing complexity.

Frequency-Dependence in the Rainforest

Frequency-dependent selective pressures help to achieve a balance among different alleles at the H locus in the gene pools of *P. dardanus*. Such equilibria add to the capacity of this species, and many other species, for future evolutionary change. And now there is growing evidence that frequency-dependent selection by predators, pathogens, and parasites also maintains and increases the number of species in entire ecosystems.

The first theories that addressed this possibility were proposed at the beginning of the 1970s. Working independently, Daniel Janzen of the University of Michigan and Joseph Connell of the University of California at Santa Barbara addressed the question of why there are so many species in a complex ecosystem such as a rainforest [10, 11].

Imagine a tree of a given species, growing in a forest. Over time this tree will attract a variety of fungi, bacteria, and viruses. Some of these pathogens will be specialized to attack the roots and leaves of this particular tree species. A roistering crowd of equally specialized insects will also arrive to munch on the tree's leaves and burrow around under its bark. This horde of parasites might have killed the tree when it was a seedling, but by the time the parasites build up to dangerous numbers the tree is already large enough to withstand their damage.

Even though the tree is able to survive, it has trouble reproducing. Most of its seeds fall nearby, which means that they are killed by the crowd of specialized pathogens and parasites that are also attacking their mother. But birds and mammals that love the seeds of this species will also gather round the parental tree. Some of its seeds will be carried, by these animals or by the wind, to more distant places. There, free of parasites and pathogens, the seeds will be able to sprout and grow.

Such a scenario should result in a forest in which all tree species are *over-dispersed*, because young trees can only survive if they are not near their parents. As a result you might have to walk for a long distance through the forest to find two trees of the same species. It is just such a situation that Charles Darwin's contemporary Alfred Russel Wallace described in 1878 as he recalled his treks through Malaysian forests:

> If the traveller notices a particular species and wishes to find more like it, he may often turn his eyes in vain in every direction. Trees of varied forms, dimensions, and colours are around him, but he rarely sees anyone of them repeated. Time after time he goes toward a tree which looks like the one he seeks, but a closer examination proves it to be distinct. He may at length, perhaps, meet with a second specimen half a mile off, or may fail altogether, till on another occasion he stumbles on one by accident [12].

Such over-dispersal, Janzen and Connell proposed, depends on the presence in the forest of pathogens, parasites, and seed-predators that are specialized to prey on particular tree species. If there is such a highly specialized component to the invisible world, the older trees' burden of parasites will kill their young. Young seedlings of a tree species are more likely to survive in parts of the forest where adult trees of the same species are rare, and where there is no gang of parasites waiting for them (Figure 18).

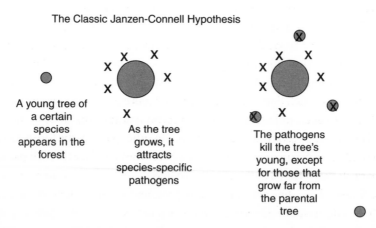

The Classic Janzen-Connell Hypothesis

A young tree of a certain species appears in the forest

As the tree grows, it attracts species-specific pathogens

The pathogens kill the tree's young, except for those that grow far from the parental tree

Figure 18 The original Janzen–Connell model predicts that trees of all species should be scattered quite evenly through the forest.

If such frequency-dependent interactions are common, the forest should reach an equilibrium state that is made up of many tree species. Specialized pathogens prevent a species from becoming locally abundant, but trees get a boost in survival if they are surrounded by other species and their pathogens have not yet tracked them down. Such a diverse forest would have truly reached a green equilibrium, in both the metaphorical and the physical sense. But does this scenario really happen?

In 1994, I was able to make a contribution to this question. In August of that year I visited the astoundingly diverse rainforest in Peru's Manu National Park, and wondered whether it would be possible to detect the Janzen–Connell effect. I knew about some wonderful sets of census data that had been collected in rainforests in Panama and elsewhere by ecologists Stephen Hubbell, Robin Foster, and Richard Condit, along with their teams of dedicated students and volunteers. These censuses were more ambitious than any censuses of the natural world that had been carried out up to that time. The censuses covered areas of pristine forest half a square kilometer in area, and they kept track of all the trees with a diameter greater than one centimeter. A single census often involved the identification and measurement of over half a million trees. Some of the plots have been censused repeatedly.

When Hubbell and his colleagues began to analyze the earliest sets of these data, from a forest on an island in Panama, they immediately discovered that the trees of each species in the forest plots tended to be clustered instead of over-dispersed. They thought that this clustering reduced the likelihood that the Janzen–Connell effect was operating in the forest, because the Janzen–Connell prediction was that the trees should be more evenly spaced [13].

But Richard Condit and I realized that clustering of species did not rule out the Janzen–Connell model. If a cluster of young trees could survive in the forest, and if it took a little while for their specialized pathogens and predators to track them down, then the whole cluster could start life in an environment that was relatively free of such dangerous pathogens (Figure 19). Gradually, as the pathogens began to accumulate, most of the reproduction of the trees in the cluster would be through seeds that were scattered to other parts of the forest. Old clusters would die out, and new clusters elsewhere would replace them. If pathogens and parasites are limited in their ability to find new trees, and if there are regions of the forest where they are not initially present, then new clusters of seedlings should be able to survive and grow in those regions.

I used a statistical approach to compare the real Panama forest with "randomized" forests with the same clustering of species but in which recruits were scattered at random among the clustered trees. There were fewer recruits in clusters of trees in the real forest than were found in the equivalent clusters of trees in the randomized forests [14]. Most of the common species in the forest showed this frequency-dependent pattern of recruitment.

Frequency-dependent effects on tree recruitment were clearly operating in the Panama forest. Since that time we have found strong evidence for similar effects in other tropical forests [15].

Recently I reanalyzed the forest data and found more evidence for pervasive frequency-dependence, even among the rarest tree species. In the new analysis I examined the rate of growth of the trees, reasoning that frequency-dependent influences such as the effects of pathogens could slow their growth.

I divided the forests up into tiny subplots, five, ten, or twenty meters on a side. Within each subplot I separated the trees into big and small. Then I examined the smaller trees in each subplot, and divided them up into those that were of the same species as the bigger trees and those that were of different species. Finally, I determined the average growth rates of the two groups of smaller trees.

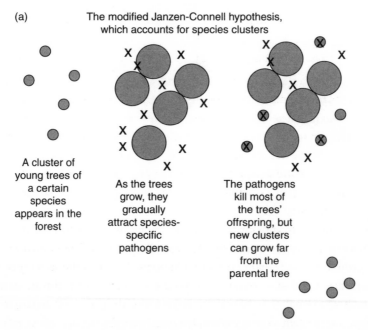

(a) The modified Janzen-Connell hypothesis, which accounts for species clusters

A cluster of young trees of a certain species appears in the forest

As the trees grow, they gradually attract species-specific pathogens

The pathogens kill most of the trees' offspring, but new clusters can grow far from the parental tree

Figure 19A Our modified Janzen–Connell hypothesis assumes that trees recruit into the forest in clusters. As pathogens that are specialized to prey on this species of tree accumulate, few young trees will appear in the cluster and new pathogen-free clusters will begin to grow at some distance from the original cluster.

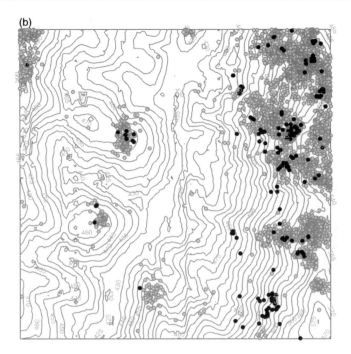

(b)

Figure 19B The black dots on this map of a forest plot in Sri Lanka show recruits of the tree *Shorea worthingtonii* during the period 1997–2007. The recruits often appeared, not in places where there were large numbers of older trees of the same species (gray dots), but on the periphery of older clusters or as totally new clusters. Data kindly provided in advance of publication by my colleagues Nimal and Savitri Gunatilleke and Suranjan Fernando.

The Janzen–Connell theory would predict that small trees that are the same species as the larger ones should grow more slowly than those of different species, because they share pathogens with the larger trees. This is exactly what I found (Figure 20). Again, I saw this effect in each of the censused rainforest plots from around the world.

When I randomized the positions of the smaller trees as a control, the differences between the two groups of smaller trees disappeared. This control showed that the growth rates of smaller trees that happened to be of the same species as the larger ones, but that were taken from different parts of the forest, were not slowed relative to small trees of other species. It was the proximity to larger trees of the same species that slowed the smaller trees' growth (Wills, in preparation).

But is the frequency-dependence that my colleagues and I detected really traceable to the effects of pathogens, parasites, and seed-predators? There could be several other possible explanations. For example, the clusters of adult trees

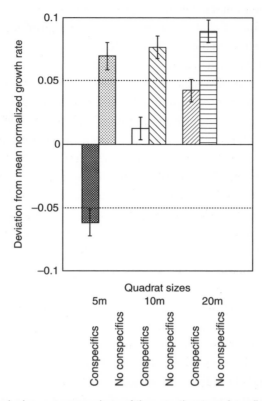

Figure 20 This graph shows a comparison of the growth rates of small trees when there are large trees of the same species nearby with the rates of small trees that do not have such *conspecific* larger trees nearby. Small trees are significantly slowed in their growth by the presence of large conspecific trees, but the effect weakens when the forest is chopped up into progressively larger quadrats. Pathogens and other frequency-and density-dependent factors would be expected to produce this pattern of growth, because their influence should be highly localized. The data are from a forest plot in Panama, but similar patterns are seen throughout the tropics.

might simply be exhausting some nutrient in the soil that the species needs. Perhaps it is nutrient depletion, not the effects of pathogens, that kills seedlings and slows the growth of those that survive.

Nutrient depletion continues to be a possibility, but direct experiments have shown that one important part of the invisible world is indeed involved in these frequency-dependent effects. Alissa Packer and her colleagues at Indiana University carried out greenhouse experiments on black cherry seedlings. They found that seedlings raised in soil that had been dug from near the roots of mature cherry trees were often killed by the bacteria and fungi that the soil contained. When soil was taken from areas where there were no mature cherry trees, it had no harmful effect on the seedlings. The harmful effects on the seedlings disappeared when the soil was sterilized [16].

A similar but more extensive experiment was carried out in 2010 by Scott Mangan at the University of Wisconsin, Milwaukee, and his colleagues. They examined six different tree species that grow in a Panamanian forest near where we had earlier discovered possible evidence of Janzen–Connell effects. As with Packer's cherry tree seedlings, soil samples that had been taken from areas close to parental trees suppressed growth in seedlings of the same species. Again, the effect disappeared when the soil was sterilized. The unsterilized soil did not suppress growth in seedlings of the other five species. Mangan's group was able to measure these effects both in greenhouse experiments and in seedlings that they planted in the trees' home forest.

Mangan and his colleagues have not yet determined which of the many bacteria and fungi in the soil are responsible for killing the seedlings. But it is only a matter of time before the agents that are responsible for these frequency-dependent effects are identified.

Pathogen-driven frequency-dependent effects may not be confined to rainforest trees. In 2010 I served on the doctoral committee of Kristen Marhaver, a coral biologist who was doing her thesis with Stuart Sandin at the Scripps Institution of Oceanography. She was studying the high diversity of corals at a coral reef on the Caribbean island of Curaçao. We talked about the possibility of trying an experiment similar to Mangan's, in order to look for frequency-dependent selection.

Kristen sampled bacteria growing on and near large heads of the brain coral *Montastraea faveolata*. She found that the bacteria inhibited the growth of young brain corals, but young corals of different species were unaffected. The effect on the brain corals disappeared when she added antibiotics to the seawater [17].

Kristen's results suggest that negative frequency-dependence of the Janzen–Connell type might be producing a green—or perhaps more appropriately a rainbow—equilibrium among the many different coral species of Curaçao. We can now begin to see a connection between the mechanisms that maintain dazzlingly diverse ecosystems such as the coral reefs of Curaçao and Sombrero Island and those that maintain the extraordinary diversities of terrestrial rainforests.

Deconstructing a Rainforest

Packer, Mangan, and others have shown that the soil of a tropical rainforest is filled with tiny organisms that help to maintain its green equilibrium. But larger creatures are likely to play a role as well. Beetles, aphids, leaf miners, caterpillars, and a legion of other insects prey on the trees. It will be much more challenging to find out whether these insects also have frequency-dependent effects on their host plants.

A first step would be to count how many of these insect species there are, and how many of them are specialized to prey on particular types of host tree.

During my most recent trip to New Guinea, in the fall of 2010, I saw first-hand the challenges that must be overcome in such a study. Through the kind efforts of Chris Dahl and Liza Bardonado of the Binatang Research Center, I was able to visit two of the Center's remote research stations.

One of these outposts, the Swire Research Station that administers the Wanang Forest Dynamics Plot, is in the hot and humid northern tropical lowlands. Like the plot in Panama it forms part of the Smithsonian Institution's planet-wide network of rainforest research plots, and we will explore it in Chapter Twelve.

The other outpost, the Ubii Camp, is located near the village of Yawan in the cloud forest of the Saruwaged Mountains. The Saruwageds are a young mountain range that lies to the east of New Guinea's main mountain backbone. Ubii, unlike the Wanang plot, is bathed in the cool clear air of the highland mountains.

At Ubii, local and foreign scientists are literally taking parts of the cloud forest apart piece by piece. A major goal of the project is to disentangle the food webs of these complex ecosystems, in order to understand in detail how they can maintain their diversity [18]. As we will see, the answers to these questions cannot be obtained without drawing on the talents and the capabilities of the people who live in the forest [19].

Getting there was a challenge. When I took off for the camp from Madang, a steamily tropical city on the northern coast, it was a little later in the morning than I had hoped. My pilots, from Canada and Australia, were both confusingly named Richard. The two Richards were doubtful about whether we could reach Yawan before the clouds moved in, but they promised to give it their best shot.

We flew east in brilliant weather along the coast of New Guinea's Huon Peninsula, with the island of New Britain clearly visible to the north. Then we turned inland and immediately found ourselves among the green canyons and knife-edged peaks of the western highlands.

The clouds grew thicker as we flew up into the mountains. Richard and Richard steered towards holes in the clouds, serenely confident that further holes would open up on the other side and they could then fly through those in turn. In New Guinea this is a sensible flight plan, because most clouds have mountains inside them.

Figure 21 Yawan's twisted and vertiginous landing strip can be seen near the top of this steep hill as we zoom in for a landing that seems guaranteed to produce a cardiac event. The profound geographic isolation of remote communities like Yawan has produced an extreme fragmentation and diversification among New Guinea's human cultures. Scientists working near Yawan are now asking whether New Guinea's jagged landscape has also had an effect on the evolution and distribution of other species.

We dived close to a mountain and dodged around it. Richard the copilot gave me a thumbs-up, and we skimmed along the valley. The minimalist Yawan airfield suddenly appeared, looking like a grass epaulette sewn crookedly onto the shoulder of the mountain. It was in the shape of a ski jump with a decided twist in the middle, resembling a Möbius strip more than a landing strip (Figure 21). We touched down on the lower part of the strip, zoomed downwards for a moment, then bounced up the slope at a gradually increasing angle. In effect we were reversing the path that a ski jumper might have taken if the slope had been covered with snow. By the time we came to a stop at the head of the strip, the plane was pointed upwards at a twenty-degree angle.

A crowd of village children and a trio of smiling guides were waiting as I emerged from the plane into the diamond-bright air of the highlands. The guides were all workers at the station—botany assistants Samuel Joseph and Nason Semi, who help to identify the many species of orchids and other plants, and porter Samuel Muwe.

We had a four-hour hike ahead of us to reach the Ubii Camp. As we followed the trail away from Yawan we saw a magnificent waterfall cascading down a wall of mountains to the east. The tops of these mountains now form part of a nature reserve for New Guinea's charismatic but desperately endangered tree kangaroos. The reserve, which was linked together in 2009 by the people of Yawan and 34 other villages in the region, is the most extensive protected region in Papua New Guinea. The agreement among the villages was negotiated with the help of Conservation International and researchers at Seattle's Woodland Park Zoo.

Radio-tagging shows that the kangaroos habitually travel more than a hundred kilometers in a season, moving back and forth between the mountain tops and isolated parts of the northern coast. Alas, a visit to the tree kangaroos' mountaintops would have taken at least a week, with no guarantee of a sighting of the elusive creatures. The kangaroos would have to wait for another trip. Instead we headed straight up the mountain on the western side, crossed its knife-edge crest, and scrambled down a wet and muddy slope into a deep valley.

Several streams trickled down the mountainsides and came together to flow across a little bit of flat land at the bottom of the valley. As we descended we passed through a patchwork of cleared fields. People from the village are beginning to hike in to the valley, in order to clear land and grow taro and sweet potatoes.

Vojtech Novotny, an ecologist from the Czech Republic, has astutely taken advantage of this clearing activity. Novotny has spent much of his working life in New Guinea. In 1997 he established Madang's Binatang Center, one of the island's first non-governmental research institutes, to carry out ecological research.

Novotny persuaded some of the farmers to let his team clear some of the land for them. He offered to do the clearing for free—though he was at pains to point out that it would take a bit longer than if the farmers were simply to wade in with their machetes.

When I arrived at the camp it was obvious why the clearing was taking so long. Novotny's team was carefully deconstructing two one-hectare plots of forest, one that had never been cut and one that had been cleared some years earlier and that had been taken over by second growth. They chopped up each tree and bush into fragments. Each insect that emerged from the fragments was isolated in its own clear plastic bag or vial, in order to identify it and observe the rest of its life cycle (Figure 22).

Novotny's team consists primarily of parataxonomists, local people who have received on-the-job training as scientists and observers.[2] Some of the team members are only one generation removed from life in Stone Age villages. They watch carefully as leaf miners, beetles, caterpillars, flies, and wasps crawl out of each leaf and bit of wood. Their observations, double-checked by taxonomists

Figure 22   Each bag and vial in this workshed at the Ubii Camp contains a fragment of tree or bush taken from a pair of one-hectare plots of highland forest. The insects that emerge from each of these bits of plant are carefully identified, counted, and sorted.

under the leadership of Scott Miller from the Smithsonian Institution, are helping to construct a complete three-dimensional picture of the plant and insect diversity in the forest. And the team is also discovering which of these insect species are specialists that dine on only one host plant, and which are generalists that are capable of chowing down on a variety of hosts.

Sometimes there are jaw-dropping surprises. On my first morning at the research station, parataxonomists Kenneth Molem and John Auga showed me—with a flourish—one of the most astonishing creatures I have ever seen (Plate 9). They had found this blazing, flame-like caterpillar during their deconstruction of the virgin forest plot.

The caterpillar was one of hundreds that the station's workers had carefully placed in plastic bags, along with what they hoped would be tempting food plants. Alas, as I learned later, this gorgeous creature died before it could metamorphose into an adult. If it had lived, it would have become an Anthelid lappet moth, one of a group of large and cryptically-colored Austronesian moths. Another specimen of this rare caterpillar will have to be found and raised to adulthood before it will be possible to determine its genus and species.

Generalists and Specialists

One big question that these surveys can begin to address is whether insects are actually helping to maintain the diversity of the forest through negative frequency-dependent selection, in the same way that soil bacteria and fungi do. Such a frequency-dependent effect of insects on their hosts could happen in two ways.

First, there may be some generalist insects that can, like top predators, switch their attention from one host tree species to another. Whenever one of these insects' host tree species becomes locally common, its vulnerable saplings and seedlings will be more heavily infested and be more likely to die before they reach adulthood. It is this kind of frequency-dependence that is commonly harnessed by gardeners and farmers, who plant tempting "trap crops" such as marigolds or Indian mustard in order to lure insects away from their real crop plants. Some American cotton farmers even plant rows of corn in their cotton fields to attract *Heliothis* moths.

Second, there may be specialist insects that can only survive, or inflict most of their damage, on one host. These specialists would rise and fall in numbers along with their host species.

In fact, the web of associations between host plants and the insects that prey on them in any ecosystem is incredibly complex, and the opportunities for frequency-dependent selection on both parasites and hosts are legion. Consider just a few examples that I have culled from the recent literature.

Leaf-cutter ants in the American tropics induce a chemical response in the plants that they attack most heavily. This accumulation of toxic chemicals makes the leaves less desirable for the ants' fungus gardens, and when this happens the leaf-cutters switch their attention to other plant species [20]. Christmas beetles in northern Australian rainforests pick host tree species on the basis of chemical signals. The ability of these host species to resist the beetles can change over time, depending on the intensity of the infestation [21]. In the northeastern U.S., fern moths choose to lay their eggs on two different fern species according to a complex set of parameters that includes the health and life history stage of the host plant and the number of potential enemies of the fern moth that each of the host plants harbors [22].

Cutworms in China's Shandong Province switch among four different host plants as a result of a wide variety of factors, some of which are frequency-dependent [23]. Some species of Passiflora (passion flower plants) have evolved structures on their stems that mimic the eggs of Heliconid butterflies. The butterflies will refrain from laying more than one egg on a host plant that has mimic eggs, even if the host plants are rare [24]. Myrmecophilous (ant-loving) Lycaenid butterflies prefer to lay their eggs on plants where there are already ants of certain species in residence. As their caterpillars grow, they secrete drops of nectar that feed the ants, while the ants protect the caterpillars from the wasps and other parasites that would otherwise attack them [25].

Such plant–insect interactions tend to be both complicated and highly frequency-dependent. We can get some feeling for how numerous such possible interactions must be from estimates of the huge numbers of tropical insect species but it has been a challenge to make these estimates.

In 1982, Terry Erwin of the Smithsonian Institution tented and fumigated nineteen *Luehea* trees in a lowland forest in Panama. He used canvas sheets to catch the insects—mostly beetles—that showered down to the ground. He ended up with about a thousand species of beetles and two hundred species of other insects [26].

Erwin guesstimated that thirteen percent of these insects might be specialists on the *Luehea* trees. If he was right, there were about 130 species of specialist beetle that could live only on this species of tree. Since beetles make up about forty percent of all the insect species in the tropics, extrapolating from these results suggests that each tree species in a tropical forest might harbor about three hundred species of insect that could live on no other host.

At the moment we know of about fifty thousand different tree species in the world's tropical forests. This number is not likely to increase much, because tree species are relatively easy to find. So, 300 times fifty thousand would yield about fifteen million insect species that are uniquely confined to their hosts. Ervin supposed that there are an equal number of insects that are not such narrow specialists, so there could be as many as thirty million insect species living throughout the tropics.

To get to this number Erwin piled one near-guess on another, the equivalent of piling Pelion on Ossa. Unsurprisingly, though everyone agrees that there probably are millions of insect species in the tropics, his huge estimate was greeted with disbelief.

Other ecologists have arrived at much lower but still substantial numbers. Novotny carried out extensive surveys in several small plots of lowland New Guinea forest in the late 1990s. He and his coworkers collected about 50,000 individual insects of almost 1,500 species from 224 species of plant [27]. They used all kinds of ingenious approaches to try for as complete a count as possible, including killing individual trees and then trapping the beetles and other insects that were attracted to the dead trees.

They divided the insects into 24 distinct groups, classified according to their ability to chew and suck on every part of the plants. During their census they were able to trace out 6,800 distinct links between the insect and plant species, an awe-inspiring number that nonetheless probably made up only a small fraction of the total.

Because Novotny's team examined so many plants and insects, they were able to get good estimates of the numbers and types of relationships among hosts and predators. Each tree species in their plots, they estimated, was host to about 250 insect species, of which 48 were specialists on that tree species alone. Based on this, the total number of specialist insect species would be 48 times the 50,000 tree species in the tropics. This gives a total of about two and a half million insect species that can live only on particular hosts, about a sixth of Erwin's estimate but still pretty astronomical.

Thus there seem to be plenty of such insect specialists, of the type that might be important in frequency-dependent interactions. But how many generalist insects are there? Generalist insects can also take part in frequency-dependent interactions if they are able to modify their search images and home in on the commonest tree species.

There are four times as many generalist as specialist insect species in Novotny's bits of forest, but the generalists may range widely because they are not tied to particular host species. This means that approximately the same set of generalists might be found in another part of the forest, even though it has a different set of host trees. And if many of the generalists turn out to have really wide distributions, then the total number of insect generalists may be substantially smaller than the total number of insect specialists.

There are lots of guesses in these estimates. The only way to settle the question would be to do thorough deconstructions of pieces of virgin forest in extremely different parts of a tropical forest, where the specialists are likely to be different but the generalists might be shared. This is what Novotny's group of parataxonomists is starting to do in the highland forest near Yawan, and it may lead to a much clearer idea, not only of the total number of insect species, but also of the relative numbers of generalists and specialists.

As the surveys continue, estimates of the number of insect species will continue to fluctuate. My guess is that the total is likely to be larger than Novotny's estimate and smaller than Erwin's, perhaps six million or so. But even if there are a "mere" six million insect species in the tropics worldwide, these provide plentiful opportunities for a mind-boggling array of trophic interactions.

Many of these interactions are sure to be frequency-dependent, probably in ways that are much more complicated and interesting than those suggested in the original Janzen–Connell model. But all of these interactions will contribute to the green equilibria of tropical forests, and perhaps of other ecosystems as well.

The Preservation of Diversity and Green Equilibria

With regard to the conservation of ecological diversity, the news from these preliminary studies of the rainforests is a mixed bag. The work of Novotny's team and of other tropical ecologists suggests that a holocaust of unknown species is not yet taking place. Many tropical insect and tree species tend to be relatively widely distributed, which means that when a small piece of rainforest is destroyed it is unlikely that its loss will drive an entire species of insects or plants to extinction. These findings give us, perhaps, a little breathing room. But if rainforest destruction becomes so widespread that entire tree species are lost to extinction, each of these extinctions will take dozens of unique insect species with it, including species as yet unknown to science.

Evolution and ecology have combined to produce millions of trophic interactions among plants, insects, birds, and mammals, and even more interactions with a vast zoo of invisible bacteria, fungi, and viruses that operate for the most part below the radar of our perceptions. The result has been stable, robust ecosystems in which ecological niches have multiplied. Especially in the teeming tropics, the sheer number of niches ensures a green equilibrium in which the numbers of each species are kept in balance. Many of these interactions are characterized by negative frequency-dependence. And in order to maintain the green equilibrium of an ecosystem, it is essential to retain as many as possible of these interactions, even the invisible ones.

How can we best preserve these ecological and evolutionary treasure houses? Can we simply sequester a few of them and try to preserve them through careful stewardship, while we continue to exploit the rest of the planet and destroy the green equilibria that once existed there? Or can we work with these ecological rules rather than against them, and try to preserve and perhaps enhance green equilibria even in places where our own population is the densest? In the chapters that follow I will explore these possibilities.

Let us turn now to the question of stewardship. Can it work? What are the dangers of pursuing this course?

CHAPTER THREE

Stewardship and Its Perils

$\mathbf{A}$s my wife and I trudged up a steep dirt road in Thailand's northern dry forest, looking for the turnoff that led to our camp for the night, we were startled by a roaring noise. Tigers are exceedingly rare in this region near the Burmese border, but there was always an outside chance that one might appear.

It was not, of course, a tiger. Instead it was a young man on a motorcycle, with a cage that contained a haunted-looking black and crimson oriole, *Oriolus cruentus*. Several small twitching cloth bags were dangling from his motorcycle. The bags contained dozens of struggling songbirds that he had trapped, using the oriole as a lure (Figure 23). I snapped a clandestine photo of him just before he sped down the road.

The Emptying of Thailand's Forests

A few minutes later we were hiking away from the road along the narrow trail that led to a mountaintop Akha outpost where we were to spend the night. Everywhere on the trail and on the road we found little clusters of feathers that had been plucked from hapless birds. Then another pair of bird trappers suddenly swept past us, one on a motorcycle and the other on a bicycle.

Trapping of songbirds is an ancient tradition in Thailand. Sometimes the birds are snared one day and carried to temples the next, to be released during Buddhist ceremonies. Other birds are sold in the markets as pets, but many are killed and eaten.

All of the poachers that we encountered that day carried homemade guns. One of them had an air rifle, but the other two were more formidably armed. These poachers are easily able to shoot mammals that range from the small barking deer to the wild gaur and banteng cattle, as well as critically endangered

Figure 23 This young motorcyclist has just emerged from a forest in northern Thailand, carrying the caged oriole that he uses to lure other songbirds to his snares. The Thais have traditionally trapped songbirds, but poaching of all animals throughout Thai forests is now reaching epidemic proportions.

large birds such as hornbills. All of these animals are in danger of disappearing entirely from many of Thailand's forested areas.

Vast forests still cover much of western and northern Thailand. Over the last fifty years almost a hundred national parks have been established, and logging has been banned in the forests for thirty years. In theory, then, Thailand should be a conservation success story.

But the pressures on these forest ecosystems are overwhelming. Between 1990 and 2005 about ten percent of the forests were logged illegally [1]. Poachers are everywhere, and the few rangers who try to catch them are outmanned and outgunned. Making matters worse, poachers and loggers can easily slip into the remote back country on the small trading boats that cross the Mekong River that separates northern Thailand from Laos (Figure 24), and through the unpatrolled forests along the Thai–Myanmar border.

Villagers in the highlands lead lives of grinding poverty, and the drug trade presents overwhelming temptations. Thailand's King Bhumibol Adulyadej, along with his late mother, the common-born Princess Mother Srinagarindra, established a number of demonstration projects to try to wean these ethnic

Figure 24 These boats carry poachers and illegal loggers across the Mekong between Thailand and Laos. There is also a brisk trade in drugs, animals, and even scrap metal.

minorities off the drug trade. The projects have had some success. Villagers who live near good agricultural land are encouraged to raise tropical fruit, and the markets in the northern cities of Chiang Mai and Chiang Rai are bursting with fruit.

But the more remote villages in Thailand's north are often hit by drought and have few alternatives to illegal trade. In one day-long hike we walked along a drought-seared valley and straight up a mountainside to reach the tiny Lahu village of Kup Kap (Figure 25). In microcosm, the village demonstrates why poaching and the drug trade are still rampant in this area.

Women do most of the work in the village, while the older men hide in the little houses and smoke opium.

Because of poaching and logging the green equilibria of the forests, long in the making, will be lost. And unfortunately, these illegal activities have distorted the way of life of many of the indigenous people in the area. Many of the people who are most dependent on intact ecosystems have been displaced.

Most of these events have taken place off the world's radar screen. And they have disrupted the potential for indigenous people to contribute to

Figure 25 Lahu villagers in the village of Kup Kap pound and winnow rice using age-old techniques. This village, near the Myanmar border, gains some revenue from tourism, but opium use is still widespread, especially among the men. In villages even more remote than Kup Kap, the opium trade and the trade in animal parts are essentially unchecked.

the maintenance of ecological balances in these remote ecosystems. As we traveled through northern Thailand we encountered some of these disruptions.

In northern Thailand few visitors venture beyond the beautiful temple-filled city of Chiang Mai. But one area that does attract thousands of tourists a year is a scattering of refugee villages along the Myanmar border to the northwest of Chiang Mai. Paduang tribespeople (or Kayan as many of them prefer to be known), along with many other tribal groups in the area, have fled unrest and persecution in Myanmar and now live in the villages.

Some of the Kayan women, starting at age five, follow a tradition of winding coils of gilded brass around their necks. Periodically the coils are unwound and replaced with longer and heavier coils (Figure 26). The increasing weight of the coils pushes their shoulders down and gives the illusion that their necks are lengthening. Stories abound about why this practice began, ranging from protection against slave raids to protection from tiger bites. The women themselves say they do it because it makes them more beautiful.

59

Figure 26 The heavy neck coil of this young Kayan girl, one of a few who have escaped the refugee villages along the Myanmar border to settle in a tourist village near Chiang Mai, fascinates her baby.

Health problems from the coils can develop later in life, but some of the younger women are astonishingly athletic, able to play volleyball vigorously. The practice of wearing neck coils is disappearing in Myanmar itself—during extensive travels around Myanmar's Inle Lake region in 2008 we met only one Kayan woman who still wore a coil. But in Thailand the Kayan are exploited for their tourist potential.

A few of the resettled Kayan have succeeded in leaving for New Zealand as resettled refugees, and a few others have moved to tourist villages like one that we visited near Chiang Mai (Figure 26). As the Kayan scatter, like many other displaced groups in these frontier provinces, they will be unable to draw on their resources and knowledge to establish sustainable tourism in their homeland.

The destruction of a cohesive human component in the tribal homelands leaves them vulnerable to exploitation from outside. Their situation stands in stark contrast to the organized groups of New Guinea tribes, some of whom we learned about in the last chapter. Because many of the New Guinea people still live in their homeland, they are able to fight more effectively for its integrity.

Out of Sight, Out of Mind

The effect of poaching has spread to more protected parts of the forest. In early February 2011 we traveled north from Bangkok through Thailand's central valley, through eye-stinging air from the burning rice stubble.

Our goal was Huai Kha Khaeng, an area of forest that had been set aside in 1991 as a World Heritage Site reserve. Together with the contiguous Thung Yai sanctuary, Huai Kha Khaeng covers almost six thousand square kilometers in Uthai Thani and Tak Provinces, an area as large as Utah's Zion National Park.

There are no villages in the reserve itself, but the entire protected region includes a number of villages with a total of about 4,000 people. Many more villages surround the reserve areas. Tak Province has a long common mountainous border with Myanmar to the west, so that there is a constant flow of poachers and illegal loggers into the region. Several park rangers have lost their lives trying to defend the animals that live there.

Huai Kha Khaeng is not a park in the usual sense. Visitors to the reserve are, in theory, not permitted beyond a small area near the entrance. But researchers are allowed full access. In company with graduate student Nanda Chai from the Thai Forestry Department, we drove through the visitor entry point to find the ranger outpost deserted. Unchallenged, we followed a winding dirt road into the restricted zone. After about twenty kilometers we arrived at the research station, where detailed studies of this mixed dry and riparian forest are being carried out under the direction of the Thailand Forestry Service's Sarayudh Bunyavejchewin.

The reserve lies squarely on the boundary between the Himalayan biological province and the more tropical lowland Southeast Asian region. It should be rich in animals and plants from both spheres. And indeed, this apparently isolated forest appears to be thriving. Each morning we were greeted by melodiously descending calls from troops of white-handed gibbons as they swung through the trees (Figure 27). A porcupine and a civet cat visited the station each night and enthusiastically rooted through the garbage.

The forest dynamics research plot at Huai Kha Khaeng, like the plots in Panama, New Guinea, and elsewhere, has been established through an international collaboration. It is run by the Royal Thai Forestry Department and two Thai universities, with assistance from the Smithsonian Institution's Center for Tropical Forest Science. The Huai Kha Khaeng plot is one of the plots where my colleagues and I found strong signs of the frequency-dependent processes that help to maintain tree species diversity.

Figure 27 In the Huai Kha Khaeng forest, troops of white-handed gibbons, *Hylobates lar*, still thrive. This male peers through the leaves of a *Shorea siamanensis* tree.

We walked with Sarayudh through the plot and the surrounding forest. The forest has never been logged, and in spite of poaching it still has substantial bird populations. We saw several specimens of the world's largest orchid, *Grammatophyllum speciosum*, which can weigh more than a ton. Only the massive trunks of mature forest trees can support these gigantic epiphytes.

We spent the next three days high in a tree in a wildlife hide. The hide had been built near a salt lick, where a trickling stream of mineral-rich water attracts the forest's mineral-starved birds and animals. The birds, including large flocks of vernal hanging parrots (*Loriculus vernalis*) and thick-billed green pigeons (*Treron curvirostra*), flocked around our hide in large numbers. They drank happily but nervously, flying off at the least disturbance. Mammals, however, were in short supply. Over almost thirty hours of observation we saw three sambar and two barking deer. They crept timidly out of the forest and retreated after a few quick gulps of the medicinal waters.

Nanda told us that a few years earlier he had seen large herds of wild gaur cattle congregating at the lick, and even the occasional herd of a rare and endangered cattle relative, the banteng. Packs of jackals had been regular visitors, and on one memorable occasion he had seen a tiger pacing around the periphery of the clearing.

During our visit we did see signs of some herds of animals. There were fresh elephant tracks and droppings, and dug-over areas showed where wild pigs had rooted about on the path leading to the hide. But most of the animals, it seemed, had gone.

In 1993, Sompoad Srikosamatara from Mahidol University surveyed Huai Kha Khaeng and found that there were on average a mere 1,400 kilograms of large grazing animals of all types in each square kilometer in the park [2]. A year earlier, researchers in India had surveyed a similar mixed deciduous forest at Nagarahole Park in southern India. They found ten times the mass of animals per square kilometer [3].

Nagarahole, a favorite of tourists from Bangalore and Mysore, has numerous rangers who police the area effectively. When I visited the park for a few days in 2005, I saw elephants, gaur, deer, and jackals in profusion. I even sighted a pack of the extremely rare Indian wild dogs, and glimpsed the rear end of an elusive black bear as it scooted into the underbrush.

The huge differences in animal numbers between the Indian and Thai parks cannot be traced to the type of forest, which is similar in both areas. Srikasamatara concluded that the Thai forests are being emptied by uncontrolled poaching.

The poaching continues in spite of Huai Kha Khaeng's designation as a World Heritage protected area [4]. The very designation itself is rooted in tragedy. In 1989, forester and vocal conservationist Seub Nakhasathien was made director of Huai Kha Khaeng, where one of his first acts was to dedicate a memorial to the rangers who had lost their lives to poachers. He was soon forced to buy a bullet-proof vest as protection. A year later, on 1 September 1990, depressed by government inaction and the corruption of local officials, he committed suicide. Partly as a result, the government agreed to the World Heritage Site designation in 1991.

For a few years attempts were made to police the region effectively. But interest in preservation has flagged, in part because so much of the park is off-limits. The small areas that are open to visitors offer little of interest. Huai Kha Khaeng has become a symbol of remote inaccessibility—many Thai scientists I have spoken to about it are nakedly envious that I have actually visited the reserve's interior.

Among its many other effects, poaching has reduced the food sources of the most glamorous of the forest's inhabitants, the Corbett's tiger (*Panthera tigris corbetti*).

Between February 2004 and February 2005, researchers from Thailand, India, and the U.S. set up pairs of camera traps over 211 square kilometers of Huai Kha Khaeng. Their purpose was to photograph large and elusive nocturnal animals, especially tigers. Some of the traps were set up at known tiger kill sites. During that year 124 pictures were snapped of fifteen different adult tigers and two cubs. Most of these pictures were pairs, taken by two cameras facing each other that photographed both sides of the animals and ensured accurate identification [5].

How many tigers are burning bright in Huai Kha Khaeng? Clearly, far more than even the luckiest visitor might encounter. Researchers estimate that there are about four tigers per hundred square kilometers in this western forest, a number that is comparable to those found in some heavily poached areas in India. But interpreting the camera capture data is tricky. How evenly distributed are tigers in the park? For example, could there be fewer tigers in the more distant, and more heavily poached, parts of the park that were not surveyed? If so, this would lead to inflated estimates of the number of tigers. Alternatively, were there many tigers that did not trip any cameras? These hypothetical tigers would have passed unnoticed, like feline neutrinos, through the study area.

Researchers assumed that the tigers were uniformly distributed, which meant that there might be about 120 tigers at Huai Kha Khaeng itself and perhaps more than 700 in the entire forested area that adjoins the Myanmar border. But in making their estimate they were extrapolating from one spottily protected area near the park's main ranger station. It is impossible to say with any confidence how many tigers might remain in the vast forested regions that lie beyond the study area, where poachers can largely operate unimpeded. And if most of the forests are being emptied of potential prey animals, then even the tigers that can avoid being caught by poachers may be threatened with starvation.

A few decades ago, the Thai forests were different from the strangely empty forest that we watched from our blind at Huai Kha Khaeng. They more closely resembled the fragments of heavily patrolled forest that still remain in parts of India.

The forest reserves of central and southern India currently abound in chital deer, the pretty *Axis axis*. In some of these reserves the chital makes up the major food source of tigers and leopards (Figure 28). Krithi Karanth and other far-sighted Indian conservationists, realizing that tigers and leopards need something to eat, have encouraged the growth of these potential prey populations [6]. As a result of their efforts the steep decline of tigers in India seems to have

Figure 28 These elegant chital deer, *Axis axis*, in India's Ranthambhore Park provide a food source for the few dozen tigers that remain in that small and endangered reserve. Such a resource is not available in the Thai forests, where the equivalent small deer species *Rucervus eldii* has been driven to near-extinction.

slowed in recent years. Tiger numbers are now hovering precariously at about 1,700 for the entire subcontinent [7].

The Eld's deer *Rucervus eldii*, the Thai equivalent of the chital, is gone from Huai Kha Khaeng. Indeed, it is almost extinct throughout Thailand, and highly endangered in Myanmar and Laos [8]. Its disappearance must have accelerated the loss of its predators.

Direct Manipulation of Ecosystems

As we traveled through the forests and other wild areas of Thailand, we were repeatedly struck by the fact that it is the regions that have been kept away from the public that are suffering some of the worst effects of poaching. Because of their isolation, these areas have also received less funding and attention from park managers.

Some of the most heavily visited parks, in contrast, have adequate populations of Thailand's more spectacular animals and birds. But even these species

Figure 29 This great hornbill male, *Buceros bicornis*, is feeding its mate and their young in a tree hollow. The hollow has been modified by Khao Yai park rangers so that the female is better protected from predators.

have had to be nursed through brushes with extinction, sometimes through dramatic interventions by scientists and park rangers. These interventions are most effective when they are done in collaboration with the people who know the environment best.

Khao Yai National Park, one of the most popular parks in Thailand, is only 150 kilometers from teeming Bangkok. It supports a diverse avian population, including some of the world's most magnificent birds, the great hornbills *Buceros bicornis*. These birds are mostly fruit-eaters, though they will also dine on mice and insects. And they have a complex reproductive cycle that absolutely depends on an undisturbed ecosystem.

The female lays her eggs in a large cavity in the trunk of a tree. The male then seals his partner into the nest by building a wall from droppings and mud. He feeds the female through a small hole. When the chicks have hatched, the parents tear down the protective wall and the mother emerges. Then, with the help of the chicks, the parents reseal the hole. They continue to feed the safely sealed-in chicks until they have fledged.

This rather baroque reproductive method only works if suitably large and intact tree cavities are available. Many nesting attempts fail because the cavities are so large that the birds cannot wall them off successfully. University and park researchers have tried to aid the birds by modifying some of the cavities at Khao Yai [9].

The male in Figure 29 is feeding his mate and chicks in such a modified nest. The rangers have nailed a board over the opening. Now the rangers are faced with the challenge of removing the board twice, first to release the female and then to liberate her young. All these manipulations have to be carried out without disturbing the birds so much that they abandon the nest.

Nest modifications have not been sufficient by themselves to halt a decline in the park's hornbill populations. The rangers have found that the most effective method is to pay local families small sums to adopt a hornbill family and to protect it from poachers. The money for the adoptions comes from visitors to the park, and from a highly effective advertising program that encourages city-dwellers and foreign visitors to pay for the protection of specific nests. The researchers report that during a five-year period, starting in 1998, these subsidies supported over five hundred cavity-years of adoptions. It is these interventions, more than any other, that have helped to slow a precipitous decline in hornbill numbers.

Our experiences in Thailand showed us that remoteness itself is a factor that can contribute to ecosystem destruction. More attention must be paid to the world's remote ecosystems—simply announcing that they are protected is not enough. And the people who live in and draw sustenance from these ecosystems must be encouraged to become part of the process of saving them. The displaced northern tribespeople of Thailand and Myanmar are in danger of losing their connection to a natural world that they have shaped, and that has shaped their cultures. It is essential that they be permitted to re-establish this connection. Only then can they help the rest of us to understand the green equilibria that keep their world in balance.

Protecting Oceanic Green Equilibria

We saw in the last chapter how the removal of top predators may have begun to destabilize Anilao's Sombrero Island reef community. Do such marine ecosystems resemble the remote forests of Thailand, which are more endangered by a lack of policing than by human use even when that use is quite heavy? In the Philippines the answer to this question is a firm yes.

A network of local fishermen and villagers has protected the reefs of the Anilao area since 1991. They have been aided by the Haribon Foundation and the Philippines branch of the World Wildlife Fund. As the local people take up this new role, one result is a kind of frequency-dependence in their behavior patterns. Many of the fishermen are abandoning fishing and adapting their outrigger fishing boats to take SCUBA divers to and from the reefs.[1] Their search image has, in effect, been changed from hunting for fish to hunting for the best and most unspoiled reefs. As a result of this shift, coral cover and fish diversity have increased since the reserves were established.

The great strength of the Anilao reefs, and the source of their resilience, lies in their immense underlying diversity and fecundity. Their hard and soft corals are among the most diverse in the world. Because fishing is prohibited over much of the area, and the dive resorts and villagers have come together to police any infractions, there is some possibility that top predators will return in sufficient numbers and variety to restore a robust balance to the ecosystem. But even in their absence, the sheer diversity of ecological niches on the reef will keep the number of species high for a while.

The complexity of the green equilibria that are maintaining the Anilao reef communities became clear to me during several dives on a site called Twin Rocks. The site is near a number of diving resorts, which means that the corals close to the shore have largely been destroyed by the trampling fins of generations of beginner divers. But a short distance from shore one of the richest reefs in the Philippines continues to thrive.

The reef is home to a tornado-shaped school of big-eyed jacks, which often hovers over it for hours (Plate 10). Although the school would normally attract the attention of large pelagic predators, it is now free to circle unmolested and suck up the young of many of the reef fish. How long can this tornado of fish go on spinning before it has an impact on reef diversity? That will depend on how quickly the top predators can return.

It is clear that, even in the absence of the top predators, the reef is still able to support a web of interactions among its species. As I was about to take a picture of the spinning school (Plate 10), I was suddenly surrounded by a crowd of small reef fish, dominated by blacklip butterflyfish *Chaetodon kleinii*. Caught up in the perfect storm of fish on this abundant reef, I could appreciate how dazzled the nymph Danaë must have been when Zeus disguised his wicked intentions by appearing to her as a shower of gold.

PLATE 1 Chief Wilem of Wasilimo Village in New Guinea's western highlands.

PLATE 2 An immature *Melanoplus devastator* grasshopper.

PLATE 3 Black caiman on Guyana's Rupununi River.

PLATE 4 Kaieteur Falls and a tank bromeliad.

PLATE 5 *Victoria amazonica* water lilies in Guyana.

PLATE 6   The rare Guianan cock-of-the-rock.

PLATE 7 Scalefin anthias, *Pseudanthia squamipinnis*, on a Philippine reef.

PLATE 8 Mimicry in the African swallowtail *Papilio dardanus*, © James Mallet and Bernard.

PLATE 9 Caterpillar of an unknown moth species in New Guinea's highlands.

PLATE 10 Big-eye jacks on a Philippine reef.

Interlocking Webs of Life Confer Ecological Stability

As we saw during our explorations of the Guyana rainforest, the diversity of the Anilao reefs depends on an interlocking network of ecological niches. This network provides stability even in heavily impacted parts of the reef. And in some cases we may have added to the complexity.

Although top predators are essential for the balance of entire ecosystems, even in their absence rich ecosystems such as coral reefs can still support clusters of ecological niches at lower trophic levels. Because the reefs' food webs are so complex, there is room for many predator–prey interactions among less formidable predators and less noticeable prey. These interactions are maintained through their own sets of green equilibria.

In 1910 Claude Debussy wrote a prelude for piano, *The Sunken Cathedral*. Its haunting sonorities evoked the lost cathedral of Ys, which is said to emerge periodically from the ocean off the coast of Brittany and then sink again into the green depths. There is no sunken cathedral at Anilao, but not far from Twin Rocks we were able to visit a sunken casino.

Sunken casinos, of course, do not have quite the same mythical resonance as the mermaid-haunted naves of Ys. This particular casino began life as a large boat, modified for gambling and anchored close to shore. Almost forty years ago, a storm broke the boat loose from its dock and it foundered.

After a series of salvage operations, all that remains of the casino boat is a drowned lattice of girders that once framed its superstructure. I strongly suspect that the casino had been a rather sleazy enterprise in its former life, but now its colorful organism-encrusted girders conjure up a lost world of sparkling gaiety.

Local increases in the number of available ecological niches, like those that were fortuitously provided by the skeleton of the sunken casino, can set up self-sustaining regions of ecological complexity. The organisms that currently live on the girders are even more exposed than those on a coral reef. The animals of this new ecosystem are those that can survive and benefit from such exposure, such as small long-nosed fish that can probe the girders' relatively shallow layers of growth (Figure 30). As the casino micro-ecosystem ages and the layers of growth on the girders grow thicker, there will be more places for concealment. The casino community will move towards a greater diversity of species and become more like a mature coral reef. But the girders will eventually rust away and collapse, and the ecosystem will change again.

Figure 30   More than thirty years after Anilao's casino boat sank, its girders are covered with a thick layer of living organisms. The animals that have settled on the girders are not as varied as those that live on a mature reef. The predominant colonizers are soft corals (above left), hydrozoans (below right), and various species of tunicates (middle left, upper right). Hard corals are just beginning to spread (lower left). Fish that can easily browse the crevices of this new habitat include the longnose hawkfish *Oxycirrhites typus* (upper left) and the longnose butterflyfish *Forcipiger flavissimus* (lower left).

The casino micro-ecosystem will not have time to evolve towards a stable green equilibrium. But other micro-ecosystems have evolved in more stable environments. I made several night dives at a site called the Fish Market, an eponymously named area that lies just offshore from Anilao's actual fish market. On each dive I spent more than an hour cruising a few inches above open sandy areas and beds of lush eelgrass. My lights revealed unexpected riches in what seemed at first to be a rather dull environment.

Unlike the casino ecosystem, eelgrass meadow communities have a long evolutionary history. The animals shown in Plate 11 come from four of the major phyla (chordates, arthropods, echinoderms, and mollusks), and represent a billion years of divergent evolutionary history. But all of the species from these diverse phyla that live at Fish Market have converged on colors and patterns that match them superbly to the site's eelgrass and sand. As long as they do not stray too far from the eelgrass beds, these disguises will protect them from predators.

The organisms that live in this eelgrass ecosystem are far more fine-tuned to their environment than the higgledy-piggledy collection of creatures that have settled recently on the girders of the sunken casino. But both the recent casino-girder community and the ancient eelgrass community consist of interlocking, frequency-dependent webs that link small predators, their even smaller prey, and the invisible parasites and pathogens that prey on both.

A Deadly Synergy

The diversity of Anilao's underwater ecosystems has cushioned them against human impact. Any damage has a chance of being reversed, because of the new-found determination of the local communities to protect this dazzling underwater world. But the devastating effects of human exploitation are far more obvious in some reefs that are further from urban and farming areas.

From Anilao we flew to Coron Bay and sailed on a dive boat north to the middle of the Mindoro Strait, between the main Philippine archipelago and the more isolated Cuyo Islands. Our goal was the Apo Reef.

This reef, 34 kilometers long, is generally agreed to be one of the world's most diverse marine ecosystems. It was badly damaged during decades of dynamite fishing, but there have been substantial efforts at protection since 1985. The condition of the reef was monitored through the 1990s, and it showed encouraging signs of recovery [10].

Then, in May of 2006, the Philippines was hit by Supertyphoon Caloy. The storm caused widespread destruction and scores of deaths throughout the archipelago. Twenty to thirty percent of the branching corals and table corals at Apo were destroyed by this typhoon alone, and one of the two boats that were responsible for patrolling the 27,000-hectare protected area was wrecked. John Manuel, one of the monitors who surveyed the damage, reported: "Many stands of beautiful branching corals were turned into rubble, as if struck by a powerful explosion; and most of the table corals were either broken or overturned. The corals that remained standing looked dead to me because many were blanketed in white sand." [11]

When we arrived in February 2011, damage from Caloy and from other storms was still obvious. The weather was unsettled during our visit, and it was difficult to find sheltered spots from which to start our dives. But we did spend two and a half days exploring the southern end of the reef.

Most of the shallower parts of the reef were either rubble fields (Figure 31) or were covered with fast-growing, pale, and wraith-like *Xenia* soft corals. The

Figure 31 Scenes like this are typical of the shallow parts of Apo Reef in the Philippines, the result of repeated storms that have hit the already weakened reef. You can see some new coral growth in the foreground of the picture, but much of the reef that lies beyond is broken rubble, eerily empty of fish.

majority of the fish that we saw were in schools that had gathered near the vertical walls surrounding the reef (Figure 32). The most sheltered parts of these walls still harbored some intact corals.

The well-preserved reefs at Anilao (and, from all reports, the equally rich reefs south of Apo at Dumaguete) are sheltered from storms by surrounding islands. These reefs seem to be robust enough to survive intensive tourism and some agricultural runoff. But Apo Reef, a huge block of solid coral that has been building up for the last thirty million years, is used to the roughest weather. It has repeatedly been pounded by typhoons even more fierce than Typhoon Caloy. So why have Caloy and other recent storms damaged it so severely? This ecosystem's vulnerability may have begun with human mistreatment.

For more than a century Philippine fishermen have been blowing up their reefs, first with dynamite and more recently with cheaper explosives made from nitrates. After each detonation only five or ten percent of the fish that have been killed float to the surface, but as long as there are new intact reefs to destroy the fishermen are able to scoop up enough fish to make the slaughter worthwhile.

Figure 32 A school of bannerfish, *Heniochus diphreutes*, cruises past the top of a sheltered walled canyon at Apo Reef. This scene gives some idea of a former abundance that has been largely lost.

At the same time, conventional seine and trawl fishing have decimated the populations of large fish. On top of this destruction, large areas of the Philippine reefs have been poisoned with cyanide in order to capture fish for the tropical aquarium industry.

Tragically, dynamite fishing continues, especially in Luzon to the north and in Zamboanga and other parts of the southern Philippines. Catherine Demesa, who is one of a handful of Filipino ecologists trained in conservation ecology, told me that the fishermen get a $20 return in dead fish for every $2 bottle of explosive nitrate that they use.

Such destruction, while horrific, need not be a death-knell for the reefs. When the system of protected reef reserves, including Apo, was set up in the 1980s and 1990s, surveys showed that fish and coral populations still retained their species diversity even though many species were reduced in numbers [10, 12, 13].

But the damage has nonetheless been severe. As I swam over hundreds of meters of damaged reef at Apo, I wondered whether there could also be a destructive synergy between natural forces and damage by humans. One such interaction could have been set in motion by the explosions that have pounded the Philippine reefs.

In *The Darwinian Tourist* I told the story of how I was hammered by a 5.4 earthquake during a dive off the central Pacific island of Yap. The earthquake, which was hardly felt on land, was so violent underwater that my first thought was that my SCUBA tank had ruptured. The shocks from the earthquake triggered underwater landslides, one of which swept down a slope towards our little group of divers. And the shocks also broke off large pieces of coral from many parts of the reef (Figure 33). Because water is eight hundred times as dense as air, the compression waves that are produced by earthquakes or explosions can be as damaging as sledgehammer blows.

Corals have a fragile limestone structure, built up from the accumulating dead skeletons of generations of the tiny polyps that make up these colonial organisms. The explosions set off by the fishermen must have cracked the skeletons of the Apo Reef, just as the earthquake I experienced cracked the reef structure on Yap. Such hidden damage cannot be healed by the polyps, because the cracks extend far beneath the corals' living layers.

Bcause the structure of Apo Reef has been weakened by explosions, the coral will be more easily smashed into fragments by subsequent storms. Then, as the fragments are tossed around in the storm surge, they pound the remaining corals to bits, multiplying the damage. This may be why the scientists who monitor

Figure 33 This chunk of living coral, two meters across, broke off and rolled down a slope on Yap's fringing reef as a result of a 5.4 earthquake in February of 2008. (Photo from Diane Corbière, one of my companions on our dive during the earthquake.)

the reef have been so amazed by the amount of damage that has been caused by recent storms.

Changing Attitudes Towards Green Equilibria

Once the damage to Apo Reef reached a certain point, the network of frequency-dependent interactions on which the reef depended broke down. On our visit we saw only three white-tipped sharks, cruising deep in the reef's protected canyons. The top predators were no longer controlling the reef, and the smaller predator–prey and host–pathogen interactions were also disrupted by the damage to the corals. Fast-growing organisms like the *Xenia* soft corals had taken over large areas of the reef and smothered them.

Some ecosystems, like the rainforest in the interior of Guyana, have temporarily escaped the impact of human population growth, which is why they have preserved so much of their biological history. Others, like the Apo Reef and the forests of northern Thailand, have been badly damaged. Still others, like the reefs of Anilao, are under pressure but have not yet reached a tipping point. They have survived a substantial amount of human impact through a happy combination of human and environmental factors.

In much of the Philippines the danger continues. After my visit I contacted Demesa, a Program Officer for the RARE Organization who has recently introduced an ecology curriculum into the schools of her home region in Camarines Sur in the southern Luzon. She has created a mascot, a red grouper fish named Agcaton, to help bring the conservation message to local schoolchildren. She shared with me the heartening tale of Manoy, a villager who had been a dynamite fisherman for thirty years. He is now a fish warden of the new marine reserve and has become a fierce protector of the reefs. Manoy told her that before the reserve was established the villagers heard explosions ten times a day. Now, because of Rare's Pride campaign, there are no more explosions.

In both land and marine ecosystems remoteness is often a curse rather than a protection, because remote areas cannot be easily patrolled. But the success of ecosystem protection also depends on how rapidly our own attitudes change.

Consider how humans have treated one of South America's ecological icons. Harpy eagles, *Harpia harpyja*, are the heaviest of all birds of prey. Until recently they played an important role as top predators in tropical and subtropical ecosystems throughout Central and South America.

The ancestors of these raptors probably arrived in the Americas before the demise of the dinosaurs sixty-five million years ago. Those ancestral birds probably did not look or behave like eagles. Harpy eagles are on a deep branch of the raptor DNA family tree. It appears that their behaviors are similar to those of other eagles around the world, not because of common ancestry, but because of convergent evolution [14].

In November 2003, with the help of our guide Eduardo de Arruda, we found a breeding pair of harpy eagles in a remote dryland region of southern Brazil known as the *Serra de Araras* (land of macaws).

When we viewed this power couple through a spotter scope, we were awed by their immense and muscular claws (Figure 34). Harpy eagles often swoop down on young monkeys or sloths and seize their heads. Their claws pierce the soft skull bones of their prey, killing them instantly. Although there are persistent stories that the eagles have also carried away human babies using the same technique, these stories seem to be myths that have been used to justify killing these magnificent birds.

Figure 34 This female harpy eagle, *Harpia harpija*, one of a breeding pair in Brazil's Serra de Araras, shows off her wicked claws.

The remote canyons of the Serra de Araras have provided a refuge for a few of these top predators. Until recently, harpy eagles were also found hundreds of kilometers to the south in the Pantanal, a vast inland delta.

The Pantanal (Portuguese for "wetland") lies in the southern part of Brazil and spills over into Bolivia and Paraguay. It is fed by rivers from uplands to the west, and drained by the Paraguay River. During the rainy season this vast swamp extends over as much as 200,000 square kilometers, making it almost as large as the entire country of Guyana.

A single dirt road leads into the northern part of the Pantanal. This *Transpantaneira* road crosses dozens of creeks and swamps on rickety wooden bridges. The workers who repair the bridges are fueled by bottles of *cachaça*, the local brandy, and the bridges clearly show the effects of their libations (Figure 35).

The slapdash nature of the *Transpantaneira* has helped, for the moment, to slow what would otherwise have been uncontrolled drainage and development of the wetlands. As we inched our way across the drunken bridges we entered an astounding world. Plate 12 samples, clockwise from top left, some of the Pantanal's riches: A family of capybaras, *Hydrochoerus hydrochaeris*, the world's largest

Figure 35   This drunken bridge is one of dozens along the road into Brazil's vast Pantanal wetland.

rodent; highly endangered hyacinth macaws, *Anodorhynchus hyacinthinus;* the endemic marsh deer, *Blastocerus dichotomus;* and abundant caimans, *Caiman yacare.*

Like the Guyana rainforests and the Serra de Araras, the wetlands of the Pantanal are home to animals and plants that span hundreds of millions of years of South American history. But they are now being battered by human storms. The macaw populations have been decimated by the pet trade and by the spread of grazing land. Big cats are few and far between—we saw only one ocelot during our stay. And the harpy eagles, once abundant, have now disappeared.

We soon discovered why there are no more harpy eagles. Lerin Facão de Arruda, the father of our guide, owns an extensive *fazenda* that he and his son have converted into an ecolodge. He has taken the lead in protecting hyacinth macaws in the area, sometimes chasing poachers off his *fazenda* at gunpoint. But he has not always been a conservationist. He told us that some years earlier he had killed the last harpy eagle in the Pantanal, in an effort to protect his livestock from their imagined threat. He also regaled us with stories about how as a young man he had dared jaguars to attack him so that they would impale themselves on sharpened stakes in traps that he had prepared.

Now his son Eduardo has converted him, despite some initial skepticism, to the virtues of ecotourism. The money that they now make from birders and other tourists is more than the *fazenda* currently makes from cattle ranching. Eduardo too has changed his attitude towards conservation. He told us, with some embarrassment, that the reason he is so expert at finding birds for tourists is because he hunted and shot so many of them as a child.

Thanks to the recent efforts of the de Arrudas and other ranchers, some of the Pantanal's diversity has survived the impact of cattle ranching. But the primary reason that this wetland has not been overexploited is that most of it is under several feet of water during the rainy season.

Now there are plans to channel the water of the Paraguay River into shipping canals, which will remove the protection provided by the periodic floods [15]. The project, called *Hidrovia,* is on hold at the moment in Brazil, but Paraguay and Argentina have begun some dredging. The project remains a constant threat to the survival of the wetlands.

In this chapter I have only been able to cover a few of the complicated forces, human and natural, that are endangering terrestrial and marine ecosystems. Poachers have been able to loot the northern forests of Thailand, and dynamite fishermen continue to damage remote reefs of the Philippines, in part because their activities are largely ignored by the outside world. Similar grim stories, with local variations, are playing out around the world.

In contrast, the thriving reefs at Anilao, the recovery of Khao Yai's endangered hornbills, and the survival of the Pantanal's hyacinth macaws show that there can be a huge payoff if both local people and outsiders become involved in ecosystem restoration. We have already encountered the importance of synergistic interactions between species in the natural world. Now we can see how essential it is to develop such synergistic interactions between people and their environment. Only then can our species play its essential part in the maintenance and restoration of the world's green equilibria.

Even when such synergistic interactions do develop, it is often extremely difficult to heal damaged ecosystems. How much damage can an ecosystem undergo before it is unable to recover? In the next chapter we will see how the process of ecosystem recovery is governed by strict ecological and evolutionary rules. It is sometimes frustratingly hard to re-establish an ecosystem's green equilibrium, especially if that equilibrium was not robust in the first place.

CHAPTER FOUR

The Challenge of Restoration Ecology

Although I live in Southern California, I am a wimp when it comes to cold water—in particular the chilly current that carries cold water from Alaska and sweeps down California's coast. This is why, even though I live only a few hundred feet from the rich ecosystems of California's coastal waters, I tend to do my SCUBA diving in bathtub-like tropical oceans. But in July of 2010, accompanied by my brave diving companions Giovanni Paternostro and Barbara Scholz, I donned an especially thick wet suit, braced myself, and jumped into the current's choppy and frigid waters.

We wanted to see first-hand the changes that have happened to the Channel Islands Marine Reserve since its establishment in 1980 by President Carter. The Reserve protects the waters that surround a group of islands off the California coast that are now part of the Channel Islands National Park. And protection is sorely needed.

A Recovering Marine Ecosystem

Human pressures on the resources of California's coast have been unrelenting. Most famously, overfishing has destroyed—it seems permanently—the immense sardine fishing industry that John Steinbeck documented in his novel *Cannery Row*. The book, published in 1945, was a nostalgic re-creation of the lives of the people who had exploited, and eventually decimated, the sardine fishery.

As a result of unrelenting overfishing, the number of California's sardines declined by a factor of 500 from the 1930s to the 1960s. Their population has now crept up about five-fold from its low, but so unfortunately has the size of the catch that is still permitted [1].

Many other California fish species have also been overexploited, and the expanding system of protected marine reserves has been designed to protect the

Figure 36 A sheephead, *Semicossyphus pulcher*, cruises through a kelp forest off the coast of California's Anacapa Island. Such large and delicious fish are among the first to fall victim to human predation. But if they are permitted to multiply once more, can the ecosystem revert to its undamaged past?

fish that remain. Is it possible to predict, from what we have learned so far about how ecosystems work, how effective these protections will turn out to be?

Do the ecological rules that we saw operating in the Sedgwick Reserve, the Philippine reefs, and the Thai forests also apply to California's overexploited marine ecosystems? How much are the fish populations of the Channel Islands Reserve influenced by the much larger areas of unprotected ocean that surround them? Are the denizens of this underwater world, like the animals and plants of the Sedgwick Reserve, locked into the rules of island biogeography and doomed to move towards some new and less diverse ecological equilibrium? Or is it possible that different ecological rules govern these marine organisms, and will lead them to a different fate?

The initial challenge was to get beneath the waves in order to take a look. I loaded the pockets of my buoyancy compensator vest with eleven kilograms of lead weights, five more than I normally use, in order to overcome the added buoyancy of my thick wetsuit.

We had picked what we thought would be a warm week in July for our dives, but the summer of 2010 turned out to be the coldest along the Southern

California coast in thirty years. World wide, 2010 was in a tie with 2005 for the title of warmest year on record, but the local coastal conditions provided a vivid reminder of the difference between climate and weather.

The cold water caused me to gasp as a little of it leaked into gaps between the wet suit and my quivering epidermis. But worse was to come. When my companions and I swam down about 10 meters we hit the thermocline, which was clearly visible as a shimmering layer where the frigid waters above mixed with the even more frigid waters below. Below the thermocline the water temperature suddenly dropped more than ten degrees. Even the fish seemed to prefer the relatively warmer water near the surface. Crowds of them danced above us in rays of sunlight.

In spite of our discomfort the scenes that greeted us, both above and below the thermocline, were lively and healthy. It is an otherworldly experience to swim through the towering columns of brown algae that make up the peaceful gloom of a kelp forest (Figure 36). This world was much closer in spirit to the sunken cathedral of Ys than to the ruins of the Anilao casino that we encountered in the last chapter.

Rockfish, sea slugs, and sea hares were common on the bottom. Lobsters, which had been rare before the reserve was established, swarmed in the submerged caves nearer the shore. This abundance is remarkable, because the California Department of Fish and Game still permits limited fishing even in the protected areas. Many lobster traps are set around the islands during the winter trapping season, and fishing for the great schools of squid continues uninterrupted. In spite of this, the lobsters near the island are increasing in numbers and squid populations, for reasons that are not understood, are actually exploding.

The health of the kelp forest indicates that the reserves may be starting to work. Perhaps most critically, from the standpoint of overall ecosystem health, during our three days of diving we saw several large black sea bass. One of them appeared to be well over a meter long, even after discounting a diver's tendency to exaggerate (Figure 37).

These magnificent fish, along with their rarer relatives the white sea bass, form an important part of the highest trophic layers in this ecosystem. They share this task with sea lions, seals, porpoises, and the occasional large shark. Together these top predators are far more varied and numerous than the coyotes, hunting cats and birds of prey that still survive at the Sedgwick Reserve.

We have seen how the healthy diversity of a terrestrial ecosystem depends on the presence of a variety of top predators. This rule holds in marine ecosystems

Figure 37   Black sea bass, *Stereolepis gigas*, are among the top predators in the kelp forest. This is my best picture of one of these fast-moving fish!

as well. Benjamin Halpern and his colleagues at U.C. Santa Barbara have shown a strong positive correlation between the diversity of top predators and the total diversity of kelp forest ecosystems [2].

Even though the black sea bass are able to capture only a small part of this marine ecosystem's total energy, they can use it efficiently and grow to remarkable size. The record stands at over two meters, with a weight of over 250 kilograms, although such monsters have not been caught for decades.

In the 1970s and 1980s the yearly catch of black sea bass plummeted from hundreds of tons to just a few tons. Although fishing for these giants of the sea was banned in California in 1981, human pressure continues. The bass still get tangled in nets, and line fishing is still permitted in Mexican waters.

Warm-Blooded and Cold-Blooded Predators

Populations of large fish such as the sea bass have bounced back in the Reserve and other Marine Protected Areas. But some species of smaller fish have stayed constant or even declined in numbers [3].

83

These declines of common prey fish might seem to be a bad sign in an ecosystem that is supposed to be returning to a healthier state. But a subtle variant of the ten percent rule of ecosystem energy flow may explain this unexpected trend.

Some of the top predators of this marine ecosystem, such as the sea basses, have evolved to be superb at conserving energy. These fish can bend the ten percent rule, altering the balance between predators and prey.

This variation on the ten percent rule can be traced to physiology and body temperature. As I swam through the kelp forests and moved up and down across the thermocline, I had to manipulate the valves of my buoyancy compensator vest continually to pump air in and out and keep myself neutral in the water. Sea basses, in contrast, can adjust their buoyancy instantly by a slight and externally invisible squeeze or relaxation of their swim bladders. This enables them to cruise almost effortlessly, saving energy for sudden dashes to capture incautious prey.

In addition, unlike warm-blooded SCUBA divers, these big predators do not have to use up energy at a frantic rate in order to keep themselves at a higher temperature than their surroundings. When my companions and I got back to the boat from our dives we were more than ready to wolf down the huge meals of barbecued chicken, baked potatoes, and pasta that awaited us. As we frantically stoked our internal fires the sea basses continued to cruise quietly in the kelp forests, their body temperatures calmly tracking the temperature of the surrounding water.

In a terrestrial ecosystem, most of the top predators are mammals or birds. Like SCUBA divers, these warm-blooded animals need a lot of energy simply to stay warm. Partly as a result, their populations tend to be small. But in a marine ecosystem the cold-blooded top predators can be more numerous, because they have lower energy requirements.

This imbalance can lead to unexpectedly high abundances of top predators in the oceans. At the undisturbed Palmyra Atoll in the central Pacific, cold-blooded sharks make up an astonishing 44 percent of the reef fish biomass [4]. Unsurprisingly, the small fish that they prey on are not as numerous as they are on other mid-Pacific reefs where humans have decimated shark populations.

Even though the reef fish at Palmyra are not plentiful, they are still diverse. Sharks are not very bright, but they are smart enough to concentrate on the most abundant of the species of fish on the reef and to shift their attention when the mix of prey species changes. Negative frequency-dependent predation by the sharks is likely to be a large factor in maintaining the diversity of their prey around that remote atoll [5, 6].

As cold-blooded top predators such as the sea bass continue to multiply in the waters off the Channel Islands Marine Reserve, this protected region may be moving in the direction of a Palmyra Atoll-like ecological balance (Figure 38). This poses a political problem. If fishermen can point to an apparent decline in rockfish and other desirable species in spite of fishing bans, they will immediately demand an end to the bans.

Can such arguments be countered? Only if we can explain to fishermen and the general public the causes of the "Palmyra Atoll effect," along with the ecological principles that underlie ecosystem productivity and species diversity. And such arguments will only be convincing if they are accompanied by evidence that the reserves are healthier and more productive than similar unprotected areas. For example, six years after the reserves were established, their spiny lobster populations increased sixfold and lobsters caught outside the reserves also increased in size and numbers [7].

Figure 38 A rockfish, *Sebastes atrovirens*, peers from the base of the kelp forest at Santa Cruz Island. Healthy species diversity, aided by the re-emergence of top predators, is characteristic of the Channel Islands Marine Reserve.

Diversity Versus Productivity

Overfishing briefly increases productivity, but this increase is not sustainable because diversity decreases at the same time. Just as the gross national product does not measure the long-term stability of a country's economy, an ecosystem's productivity does not predict its long-term stability.

The most productive ecosystems of all are *monocultures*. Monoculture wheatfields and fish farms produce enormous amounts of food, but they also produce enormous amounts of pollution from agricultural runoff and fish wastes. In addition, crowded monocultures are notoriously susceptible to runaway diseases that can spread readily—like the sea lice that swarm in Canadian salmon farms and that can escape to infect and damage wild salmon populations [8].

If the goal is simply to produce the largest mass of commercial fish over a short time-frame, then monoculture fish farming is the way to go. But if the goal is to maximize different resources that have competing requirements, such as tourism and fishing, and to ensure that these resources will be available in the future, then it is important to look at ways to retain the ecosystem's green equilibrium. A healthy ecosystem is not the one that gives us the greatest short-term benefit, but the one that will benefit us far into the future. And it is this argument for a green equilibrium that must prevail over the long term.

As I surfaced after my last dive, in a little cove on the south coast of Santa Cruz Island, I was impressed with how the Channel Islands' marine ecosystem seems to be recovering now that commercial fishing is limited and divers are no longer decimating the populations of abalones and lobsters around the islands.[1]

Ocean ecosystems are also protected by their vastness and their relative inaccessibility. The tempting pieces of real estate that surround the Sedgwick Reserve can easily be converted entirely to human use. But the ocean that surrounds California's Channel Islands cannot be subdivided into housing developments. The Channel Islands Marine Reserve's recovery has been aided by the fact that the entire ocean that bathes these islands remains rich and diverse in spite of two centuries of overfishing. This vast ecosystem will continue to harbor many secret refuges for rare species.

As I swam back to the boat for the last time, I glanced up at the sheer rock face of the south shore of Santa Cruz Island. What about the fate of the land part of this national park? Attempts to restore the balance of the islands' terrestrial ecosystems, it turns out, have been far more complicated and challenging than the restoration of their marine ecosystems. The islands pose a different set of problems. This is because the sparseness and simplicity of their web of ecological interactions makes them inherently unstable.

Recall that the theory of island biogeography predicts that small and isolated island ecosystems will tend to be species-poor. This is because an island's smallness puts all its inhabitants at increased risk of extinction. And this increased risk poses challenges to the establishment of a stable green equilibrium. The story of Santa Cruz Island illustrates the difference between a spatially limited island ecosystem and the rich and complex ecosystem of the ocean that surrounds it.

Cattle, Sheep, and Feral Pigs, Oh My!

In June 1963 I boarded a Navy PT boat at Port Hueneme, on the California coast not far from Santa Barbara. The main purpose of the trip was to ferry needed supplies—mostly cases of beer—to the naval radar station and the ranch on Santa Cruz Island. The stack of cases strapped to the boat's deck formed an excellent protective shield against the spray as we bounced across the rough waters of the Santa Barbara Channel.

Figure 39 Waves explode from the Scorpion blowhole, not far from Prisoners' Harbor where I made both my landfalls on Santa Cruz Island.

Santa Cruz, about four times the size of Manhattan, is the largest of California's Channel Islands. Its main port, Prisoner's Harbor, is actually no more than a windswept dent on the island's north coast (Figure 39). Because of the strong currents, boats can only remain docked to the harbor's little jetty by running their engines continuously.

The Spanish explorer Juan Rodriguez Cabrillo sighted Santa Cruz in 1542, and sixty years later Sebastián Vizcaino landed briefly on the island. He reported that the inhabitants were numerous, and that the men wore beards. Then the islands were left in peace until 1769, when Gaspar de Portola claimed them officially for Spain. Regular contacts with the outside world, and the first attempts to Christianize the native population, began in 1782.

The Spanish also brought part of the invisible world with them. During the latter part of the eighteenth century outbreaks of measles wreaked havoc on the Native American populations of the California mainland. Measles, mumps, and influenza quickly spread to the Chumash populations of the Channel Islands, killing many of them.

By 1816, almost all the natives of the islands who had managed to survive had been taken to the mainland. There they became laborers and indentured servants. The remaining islanders vanished from history. The mainland Chumash also nearly disappeared, reaching a low of 200 people by the beginning of the twentieth century. Since that time their numbers have begun to recover slowly.

The islands remained virtually deserted for decades. Prisoners' Harbor got its name from a boatload of Mexican prisoners who were marooned there in 1830 by their callous *Californio* captors, after towns on the mainland refused to accept them. The resourceful prisoners managed to escape from the island on homemade rafts, making a brave voyage across seventy kilometers of open sea. But they were recaptured when they reached the mainland, and their fate is uncertain.

In the mid-nineteenth century a Santa Cruz Island Company was formed by a group of San Francisco absentee-landlord speculators. In 1880 Justinian Caire, a French immigrant to San Francisco, acquired a controlling interest in the company and traveled to his newly-acquired island for the first time. He built a little church and a comfortable ranch house and planted fruit trees and vineyards in the island's sheltered central valley. Caire's descendants eked out a living until 1937, when Los Angeles businessman Edwin Stanton purchased ninety percent of the island and began to raise cattle.

My own task on Santa Cruz was to trap specimens of *Drosophila pseudoobscura*. These native American fruit flies are distant relatives of *Drosophila melanogaster*, the more famous African fruit fly that is commonly used by geneticists. My efforts were part of a decades-long project that had been started by my mentor, the evolutionary biologist Theodosius Dobzhansky.

Dobzhansky had discovered that genetic variants in the D. *pseudoobscura* populations had changed in frequency throughout the American West over a span of decades. During an early phase of his study, in the late 1930s, he had trapped a sample of flies near the Santa Cruz Island ranch house. Now he had sent me to gather some more flies from the same site, so that we could find out how they had evolved in the intervening third of a century.[2]

It took me only a few evenings to obtain a fine sample of flies. They abounded near the ranch house and swarmed eagerly to my banana-bait traps. During my idle daytime hours I was driven around the island in a jeep by the ranch foreman, Henry Duffield, who introduced me to its rather extreme ecology.

Henry had arrived on the island through a series of accidents. A rancher and polo player, he had roistered around the world with his opera-singer wife until their marriage broke up. Then, while playing polo in Mexico, he was stricken with poliomyelitis and had to spend a year in an iron lung in Mexico City. As he was recovering he met Edwin Stanton's son Carey, who invited him to come to Santa Cruz. Henry had arrived three years before me, and even though he had lost the use of his legs he had expertly taken over the management of the ranch.[3]

Henry used a jeep fitted with hand-operated controls to drive me along the island's rough roads and tracks. The scene appeared bucolic. Live oaks grew along the margins of a stream that meandered through the central valley. The south-facing slope of the valley was rather bare and eroded, but the north-facing slope was still covered with bishop pine and the unique Santa Cruz ironwood. Hereford cattle were dotted about the valley's grassy floor. In short, the island's vistas looked like paintings by Constable (Figure 40).

The peaceful scenes masked rapid environmental degradation. At the time of my 1963 visit, descendants of the original Caire family still raised Merino sheep on the eastern ten percent of the island. The sheep had become feral through neglect, and had done what sheep do so well—eating almost every plant down to ground level and leaving little but bare earth and rock.

Henry's ranch-hands had built fences to keep the sheep out of the lush central valley. This would have worked, except for the fact that the island was also full of feral pigs. The European ancestors of these pigs had been introduced to the island in the mid-nineteenth century, probably before Justinian Caire arrived.

Figure 40A and B   I took these two pictures forty-eight years apart. The first, from June 1963, shows ranch foreman Henry Duffield with his jeep at an overlook above the western end of Santa Cruz' central valley. The second picture was taken in April 2011, more than twenty years after the last of the cattle had been removed from the island. By a happy coincidence the jeep I was using was also white, and shows only small evolutionary

The pigs, assisted by sudden floods that washed away the fences' foundations, repeatedly broke through the ranch hands' Maginot Line. The sheep, lured by the greener grass on the other side, swarmed through the breaches. Working in tandem with the pigs, they had already brought many of the island's *endemic* (unique to the island) species of plant to the point of extinction. By 1963 an endemic monkeyflower had already been lost. The pigs were also eating all the acorns, preventing the island's live oaks from reproducing.

The island's 5,000 pigs were the most destabilizing element in this precarious ecosystem. During the century since their introduction they had "reverted" to some approximation of their wild European ancestors. To use more precise evolutionary terms, alleles at many of the genes in the gene pool of the island's pig population had been shifted in frequency by natural selection, so that genetic recombination was able to bring these alleles together in combinations that had not appeared for centuries. The result was a self-sufficient population of feral pigs—lean, mean, and covered with brown hair—that was well adapted to life without the aid of humans.

In spite of the damage that the pigs were doing, in 1963 the possibility that they might one day be eradicated from the island seemed a mere Sisyphean dream. Nonetheless, some of Henry's hunting dogs enthusiastically played the role of Sisyphus. They accompanied us on our jeep rides, leaping out of the jeep at the slightest scent of pig and bounding off into the brush to hunt them.

By this time the male pigs had begun to resemble wild boars, sporting substantial tusks and posing a real challenge to the dogs. An accumulation of small grave markers near the ranch house identified the final resting places of dogs that had misjudged the boars' fighting skills.

At the time of my first visit, this highly unstable interaction among the pigs, sheep, and cattle was having its largest impact on the island's fragile ecosystem. But this situation was soon to change. A year after I left the island with my sample of fruit flies, Edward Stanton died and his son Carey inherited the ranch. Carey was determined to restore the island's natural environment. In 1978 he sold much of the ranch to the Nature Conservancy. When he died at the ranch

Figure 40A and B Continued

changes from its earlier incarnation! I was not able to take my new picture from exactly the same spot as the earlier one, because several live oak trees had grown up and blocked the view. But you can see that, during the intervening half century, some bushes and trees have invaded the grasslands of the valley floor. On the slope to the left, the spread of low bushes has healed some of the more obvious scars from grazing and erosion. The scene seems little-changed, but in fact many endangered native species, such as the Santa Cruz *Dudleya* with its beautiful white flowers, are beginning to re-establish themselves.

house in 1987, the rest of his holdings passed to the Conservancy. And in 1997 the sheep-ravaged eastern end of the island was bought by the Federal government and became part of the new Channel Islands National Park. These events paved the way for an astonishing experiment in restoration ecology.

Pygmy Mammoths and Flightless Sea Ducks

Dramatic as these recent changes have been, they are only a small part of the history of the Channel Islands.

Most of the world's islands pose special problems for the species that live on them. The small islands off the coast of California are no exception. As with the Galápagos archipelago and the Hawaiian Islands, the simplicity of the Channel Islands' ecosystems puts their species at risk. To understand why their ecological equilibria are so fragile, we must look at their history.

The islands of Santa Cruz, Anacapa, Santa Rosa, and San Miguel form the tips of a submerged mountain chain that extends west from the California mainland. But this was not always the case. During the last ice age maximum, between 20,000 and 25,000 years ago, glaciers reached as far south as Montana and sea levels around the world were about 120 meters lower than they are today. At that time the northern Channel Islands were joined together into a single, much larger island, to which geologists have given the name Santarosae (Figure 41). This ancestral island was separated from the mainland of California by only about three kilometers of open water.

Santarosae was 2,300 square kilometers in area, about five times as large as the sum of today's islands. Rising sea levels had repeatedly broken it up into individual islands, but Santarosae re-emerged as the ocean retreated again. This dynamic history made the island both an ecological refuge and an evolutionary laboratory for the animals and plants that were able to cross from the mainland.

Among these immigrants were mammoths. At some point, probably during the most recent glacial period, a few of these immense animals managed to swim across the narrow channel from the mainland.

Because Santarosae's food resources were limited, the mammoths were rapidly selected for small size. Their descendants shrank down to the size of large cattle. They became pygmy mammoths—without a doubt the most oxymoronic of animals.[4] Even after their shrinkage, the pygmy mammoths were by far the largest animals on the island.

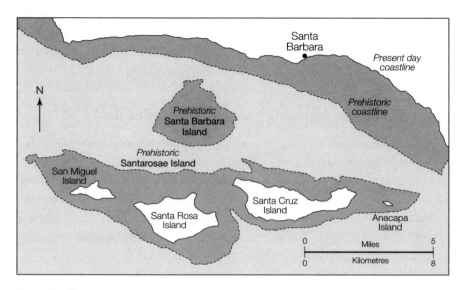

Figure 41   Twenty-five thousand years ago, at the height of the last Ice Age, the island of Santarosae was large enough to support a population of pygmy mammoths.

Mammoths were not the only unusual animals that were thriving on the islands. Numerous fossils of an extinct flightless sea duck, *Chendytes lawi*, have been found on present-day Santa Rosa Island. The oldest traces of these sea ducks go back about 40,000 years.

These birds had reduced flight muscles as well as reduced wings, and are likely to have evolved on the California coast for the same selective reasons that flightless cormorants evolved on the Galápagos Islands. Because the ancient sea ducks, like the present-day cormorants, were able to swim, dive, and feed in the rich ocean shallows, they did not need to fly. For both the sea ducks and the cormorants, big energy-gobbling flight muscles were a disadvantage [10].

The lives of the islands' pygmy mammoths and flightless sea ducks began to change when the first humans arrived. The first people who were able to escape southwards from the icebound Alaskan tundra probably used boats to hopscotch down the Pacific Coast. Crude arrows and fish-hooks, dating from about 12,000 years ago, mark the earliest signs of their arrival on the Channel Islands [11].

This date of 12,000 years puts the arrival of humans suspiciously close to the date of about 12,800 years for the most recent of the pygmy mammoth remains. And these people, or possibly later migrants, brought dogs with them [12]. Could humans and their hunting dogs have driven the mammoths to extinction? So far there is no evidence, such as would be provided by the discovery of butchered pygmy mammoth bones, to suggest that they did. Alternatively, as the sea level rose and Santarosae broke up into smaller islands, the remaining fragments of land might have become too small to support the mammoths. Or perhaps these animals were driven extinct by a combination of natural and human factors. Once the mammoths had become rare because of rising sea levels, relentless hunts and occasional kills by humans and their dogs might have been enough to wipe them out.

It is much more likely that humans were responsible for killing off the flightless sea ducks. The last of these birds disappeared about 2,400 years ago [13]. Remarkably, even though the sea ducks were probably easy prey for anybody with a club (sitting sea ducks, so to speak), they managed to coexist with humans and dogs for 9,000 years. Clearly, the island people were not simply driving all the islands' animal populations to immediate extinction. Did the islanders adapt to their environment, and learn to exploit it sustainably?

The human population was certainly thriving before the arrival of the Spanish. In 1602, Sebastián Vizcaino counted about 1,200 people living in ten different villages along the coast of Santa Cruz. This population was far higher than any that the island has since been able to support.

If native peoples had been living at this density throughout North America when Europeans arrived, their population would have totaled over a hundred million. Such a high number is unlikely, suggesting that the islanders' dense populations must have been a special case. They benefited from the sheer abundance of the marine ecosystem that surrounded them.

Nonetheless, there is evidence that they did overexploit resources such as the sea ducks and other island species. Careful examination of *middens*, the great piles of shells and bones that the people of the islands left behind, has revealed substantial changes in their diet over time. At one set of 6,000-year-old middens on San Miguel Island, the oldest layers are rich in the bones of sea otters and the shells of mussels [14]. When these resources were exhausted, the hunters abandoned the area for a hundred years. During that time red abalones were free to explode in abundance because sea otters were no longer eating them. When humans returned to the area, they switched their attention to the abalones.

These first people to live on the Channel Islands were showing frequency-dependent hunting behavior, like the top predators we have met in other ecosystems. As the abundances of their food sources changed, they repeatedly switched from one to another. Remarkably, they were able to utilize their food sources for more than ten thousand years without overexploiting them all the way to local extinction [15]. During that time, in nearby Central and South America, entire civilizations rose and fell as their peoples exhausted the land on which they farmed.

It was almost certainly the abundance of the surrounding sea that allowed the tribes on the islands to avoid the same fate as the mainland civilizations. Even when the tribespeople overexploited some species of shellfish, marine mammal, or seabird, there were so many other resources available that the overhunted populations were able to recover. Even so, some of the most vulnerable prey species were eventually driven to extinction, most notably those unlucky sea ducks.

Improvements in technology may also have increased the human impact on the islands' ecosystems. Approximately a thousand years ago the first islanders' primitive dugouts and reed boats were replaced with far more seaworthy sewn-plank canoes that the Chumash call *tomols*. This remarkable invention, which may have been introduced by the mainland Chumash tribes, increased the trade among the islands and the mainland and also increased the pressure on the islanders' food supply.

We will never know whether the island populations would eventually have crashed even without the arrival of the Europeans and their diseases. Now, the ecosystem on which they once depended is only a shadow of its former self. Even the establishment of the Marine Reserve is unlikely to restore the earlier abundance of the islands' intertidal communities.

The long precarious history of the islands illustrates the instability of its green equililbria. As we try to restore balance to the islands this instability continues to challenge us.

Restoring the Island

When I returned to the island in 2011, to see what had happened in the almost half century since my previous visit, I was met by Lyndal Laughrin, the director of the University of California's Santa Cruz Island Reserve research station. He quickly introduced me to examples of the island's current evolutionary change.

We drove through a mix of chaparral to the west end of the central valley, and switchbacked up to the crest of a ridge. There we encountered a stand of bishop pine. Oddly, the stand was a mix of vigorous young green trees and extremely dead older ones. The cones on some of the dead trees had opened, allowing their woody seeds to be released (Figure 42).

Figure 42 These bishop pine cones are scattering seeds even though the tree has been killed by a bark beetle. Because fire has been absent for decades throughout much of the island, these pines can only reproduce if they are killed by the beetles.

These dead trees, Laughrin told me, had been killed by bark beetles during a severe drought that had lasted from 1986 to 1991. In 1999, Hartmut Walter and Leila Taha of UCLA documented the paradoxical role that the beetles had played in the survival of this stand of trees [16].

Bishop pines have tightly closed cones that imprison their seeds. The cones can only open if they are dried out—through fire or, in the case of this stand of bishop pines, through the sudden death of the parent tree.

On this part of the island there had never been a recorded fire. Once the bishop pine trees had become established on the hillside, they could not have perpetuated their colony without being killed by the beetles. Like the legendary Phoenix, the trees on this hillside are only able to reproduce by dying.

Such a situation is extremely unstable. If the right combination of drought and beetle infestation does not happen in the future, and if this part of the island is not swept by fire, the new generation of trees that are growing up among their dead parents could all die from old age. Their seeds would decay inside the cones. Alternatively, given enough time, strains of bishop pine might evolve in this population that are able to open their cones and scatter their seeds without the daunting requirement of being killed first.

The fact that there are so many possible outcomes for this little colony of bishop pines illustrates the island's ecological and evolutionary instability. Whenever the environment changes in this isolated ecosystem, even in the absence of human influence, new evolutionary and ecological pressures will lead to new, precarious equilibria. When we disturb such interactions, which are already teetering on the brink of instability, it is sometimes incredibly difficult to restore them.

Nonetheless, by the time of my second visit in 2011 there had been huge advances towards a new equilibrium for the island as a whole. These were the result of a massive program of *restoration ecology*, initiated by The Nature Conservancy and the National Park Service, that was designed to remove introduced species of both animals and plants.

Plate 13 shows a glimpse of the island today. Goldfield asters, *Lasthenia californica*, bloom in a dazzling display on the island's western tip. This flat and windswept part of the island is still largely covered with introduced grasses, but with the removal of the cattle they are now beginning to yield to native bunchgrasses and to a beautiful "blue-eyed grass"—actually a native iris, *Sisyrinchium bellum*.

Once The Nature Conservancy gained control of most of the island it quickly removed the cattle. The last of them were shipped to the mainland by 1989. The feral sheep and pigs posed a far greater logistic and public-relations challenge. At the start of the removal program there were 37,000 sheep and 5,000 pigs on the island. As the trapping and slaughter began, there were fierce protests from animal rights' groups, which tried unsuccessfully in Federal court to stop the killing.

The sheep had all been trapped or killed by 2002. In an effort that was designed to defuse animal-rights protests, some of them were moved to the mainland. A small population of survivors is now maintained by the American Livestock Breeds Conservancy. Smaller but more robust than other sheep breeds, the island sheep are able to thrive under marginal conditions. Their adaptations, like those of the feral pigs, have resulted from evolutionary changes during the almost two centuries that they had lived semi-wild on the island.

Trapping could not remove the small numbers of increasingly wary pigs that ranged over the island's more inaccessible peaks. Marksmen were brought in from New Zealand, and they shot some of the remaining animals from helicopters. These marksmen had already rid several small New Zealand islands of introduced sheep and deer.

The last pigs were killed in 2007, by a combination of helicopter gunfire and the use of tethered female pigs as lures for unsuspecting males [17]. Many native plant species began to recover immediately. But the eradication of the pigs also triggered a host of unintended consequences that had their ultimate origins in the island's precarious ecological balance.

Saving the Little Foxes

All of the major Channel Islands support populations of foxes, close relatives of the mainland gray fox. These cute island foxes have become miniaturized, like the mammoths. They are now the size of large domestic cats (Figure 43). Each island fox population is distinct, and all of them are endangered.

Ecologist Dennis Murphy has called whales and pandas and other such endangered species "charismatic megafauna"[18]. His term encapsulates the

Figure 43 A and B   Above: In 1963, during my first trip to the island, I saw three Channel Island foxes, *Urocyon littoralis santacruzae*. This one was in the middle of an area near the ranch house that had recently been rooted over by wild pigs. Right: During the three days I was on Santa Cruz on my second trip in 2011 I encountered foxes nine times. This was my best picture, because it was taken in the early evening—in all the others, these largely nocturnal foxes were squinting painfully in the sunlight. Between my first and second trips the foxes had been driven almost to extinction by golden eagle predation.

Figure 43 A and B Continued

ability of these animals, through the magic of their personalities, to raise money and resources for their own preservation. Although the "mega" part of this designation hardly applies to these tiny foxes, there is no doubt that they are charismatic. And their charisma has helped them to survive a near-death experience as a species.

At the same time as the Nature Conservancy and the Park Service began to remove pigs from the island, the numbers of these charismatic foxes plummctcd. Gary Roemer of UCLA and his colleagues showed that the decline was the fault of another charismatic species, the golden eagle [19].

Golden eagles fly out to the islands from the mainland to dine on the occasional little fox. When pigs were introduced 150 years ago, they added piglets to their diet. But the golden eagles were prevented from settling permanently on the island by already-established and highly territorial bald eagles. The bald eagles eat fish and carrion, so that they are no threat to the foxes. But their ability to keep the golden eagles from nesting nearby had the collateral result of reducing golden eagle predation on the foxes and pigs.

Then, during the 1970s, this delicate balance was upset. A factory in Los Angeles began to spew DDT into the ocean, and the insecticide entered the marine food chain in massive amounts. The DDT became progressively more concentrated in plant and animal tissues as it was passed up through marine trophic levels. When the bald eagles ate the DDT-contaminated shellfish and carrion, the chemical prevented the shells of their eggs from forming properly. The bald eagles went locally extinct on the islands and almost disappeared on the mainland. The golden eagles, which dined on terrestrial animals that had much lower levels of DDT, were now free to invade the island. They quickly began to establish nests.

At first the golden eagles concentrated on the plentiful piglets. But as the restoration ecologists shot and removed the pigs, the eagles reverted to their more traditional *amuse-bouche*, baby foxes. The fox population, which was estimated to be more than 2,000 when I first visited Santa Cruz, plummeted to less than 200.

What to do? Obviously, it was out of the question to kill those charismatic golden eagles. Instead, over a six-year period, the New Zealanders and other expert hunters trapped almost fifty of the eagles and took them to the mainland. To snare the last holdouts they sometimes had to perform James Bond feats. At one point they chased an eagle with a helicopter to tire it out, and then dropped a net over it when it landed.

After the shutting of the LA factory in 1982, DDT in the environment gradually diminished. The bald eagles, safe from the pollutant that had made their eggs fragile, are re-establishing themselves. At the time of my 2011 visit there were three nests on the island.

And now the foxes are on the rebound—though not without triggering another set of cascading consequences. As the foxes increase in numbers they seem to be negatively impacting the island's endemic spotted skunks, with which they share a taste for mice and other small rodents.

It is unclear who was the first person to remark that history is just one damn thing after another. Whatever the source of this epigram, it certainly applies to restoration ecology.

Staggering Towards Equilibrium

Between my first and second visits to the island, a world that had been dominated by cattle, sheep, and pigs has been transformed into one from which all these animals have been removed. The bald eagles, which once ruled the skies over Santa Cruz, show signs of returning.

Before the arrival of humans and dogs at the end of the ice age, the foxes may have been the top mammalian predators.[5] They shared the island with pygmy mammoths, skunks, and mice. The last of the dogs on the northern islands were removed by shepherds decades ago [21]. Foxes again find themselves by default the largest mammals (aside from human researchers and tourists) on Santa Cruz. The full responsibility for being at the top of the island's food chain once more falls on their narrow shoulders. It has taken millions of dollars and the strenuous efforts of teams of ecologists, skilled hunters, and daring helicopter pilots to put them there.

After these upheavals things have calmed down somewhat. The Nature Conservancy is now turning its attention to ecological imbalances that are just as serious, but that involve less charismatic players.

Anise-flavored fennel, a mainstay of Mediterranean cooking, escaped from kitchen gardens and began to spread throughout southern California during the early nineteenth century. It had reached Santa Cruz by 1850 (Figure 44).

At first the fennel was kept under control by the cattle and sheep. The pigs also enjoyed rooting it up as they roto-tilled the land. After the animals were removed, fennel carpeted much of the island's eastern end. Extensive patches have sprung up elsewhere on the drier slopes. The flowers are briefly pretty, but

Figure 44 Gray-colored Fennel, *Foeniculum vulgare*, an import from the Mediterranean, carpets the eastern end of Santa Cruz. Such invading plants continue to threaten the island's ecological stability.

they soon shrivel into pale and sepulchral sticks that give the fennel-covered hills a leprous look throughout much of the year. And fennel is superb at smothering timid native plants.

Lyndal Laughrin thinks that because the native oaks, chokecherry, and *Ceanothus* are now free to grow and spread, they will begin to shade the fennel and bring it under control. He envisions an eventual uneasy truce between the endemic plants and the interloper. In the meantime, the National Park Service and The Nature Conservancy have experimented with controlled burning—and ended up frantically fighting runaway blazes.

The Nature Conservancy has also cleared some slopes of fennel with herbicides. These efforts have had little effect, in part because they are so underfunded compared with the massive pig and eagle programs—a total of six million dollars for the pigs and more than a million for the eagles.

More recent fights against the gray hordes of fennel involve attempts, spearheaded by a company from the mainland, to replace them with native plants. And other projects to reintroduce native plants are being led by a hardworking pair of botanists, Don Hartley and Mark Broomfield. They took me on a tour of their island greenhouses, which were filled with flats of native grasses that they are growing in sterilized imported potting soil. Some early plantings of these grasses in areas near Prisoners' Harbor look promising.

As the island teeters towards a sketchy equilibrium, a long-term problem faces the restorers of its ecosystem. Santa Cruz is small and isolated, and its trophic levels are few and sparsely populated. A pregnant female cat that escapes from a campsite, rats brought by a visiting boat, rabies or distemper from the mainland infecting the fox population, a new bark beetle or fungus—any of these could send this fragile emerging ecosystem into a tailspin.

The robust green equilibria that we encountered in the Guyana rainforest and the coastal reefs of the Philippines are able to withstand, at least in the short term, the losses of some top predators and even of large parts of their food webs. Santa Cruz, too, will gradually gain in robustness as more and more native species of plant emerge from refugia. The native plants, along with snakes, rodents, foxes, skunks, bald eagles, owls, and the occasional visiting golden eagle, will help to contribute to a green equilibrium. But the island will never move back to the more robust ecosystem that greeted the first people who landed on the island of Santarosae 12,000 years ago. The herds of pygmy mammoths, the skies filled with eagles, the immense populations of marine mammals, the great beds of shellfish along the shore, all belong to an ancient and far richer time. We can only imagine that scene, with its air filled with the trumpeting of mammoths, the cries of millions of seabirds, and the calls of great waddling crowds of flightless sea ducks.

The world has moved on. Today's Channel Islands are a remarkable testimony to the far-sightedness of organizations like The Nature Conservancy and the preservationist instincts of the people of California. But the islands are too small, too isolated, and too susceptible to drought, floods, and fire, for their new equilibrium to persist without continued human monitoring. And there are other dangers that might undo all our attempts at restoration ecology.

Existential Threats to Ecosystems

All the Earth's ecosystems are threatened, but some are more severely threatened than others. We must take these threats into account as we weigh the chances that ecosystems can be healed and their green equilibria re-established. Coral reefs provide a vivid case in point.

A visitor to the coral reefs of the Philippines can still encounter ecosystems that for richness, diversity, and sheer biological excitement are almost without peer in the world today. Even parts of the reefs that have been damaged still retain many species that are able to cling to their reduced ecological niches. But, even if we do change our behaviors, so that we foster diversity rather than diminishing it, the window of opportunity that might permit these reefs to recover is closing.

The dangers that besiege the reefs are numerous. The loss of top predators like sharks and tuna have the effect of unbalancing the trophic levels that lie below them, leading to excesses of some prey species and reduced numbers of others, as we saw in the reefs of Sombrero Island. At Apo Reef, human damage to the corals has compounded the effects of storms.

Throughout the archipelago, agricultural and industrial runoff weakens and kills the most susceptible species and permits the most resistant species to multiply. At Coron Bay, on our way to Apo Reef, I found the reefs covered with out-of-control hideous white *Synaptula* sea cucumbers. They festooned the dead corals like a plague of discarded prophylactics. Agricultural runoff into the bay and uncontrolled fishing have been much more damaging at Coron than at Anilao. The pollution has killed the sea turtles and lobsters that would normally prey on the sea cucumbers.

These factors are threatening individual reefs, but there are larger global factors that pose an existential threat to reefs throughout the world. These factors are all traceable to human-caused global warming, which leads to sea-level rise, coral bleaching, and a shift in the water chemistry itself.

The term "existential threat," which emerged in the 1980s, is usually employed by military hawks who want to frighten people with the scary capabilities of their opponents. The term is based on the original meaning of existential, which is simply the adverbial form of existence. It is important to distinguish this older meaning from the use of the word by the philosopher Søren Kierkegaard and his followers, who founded the school of existentialism in an effort to reinsert the problem of human existence into the abstract world of philosophy. I think that it is valid to use the older meaning of existential, despite the term's political baggage, to characterize threats to the continued existence of entire ecosystems.

The existential threat of global warming may already have harmed the corals throughout the Philippine archipelago and throughout the tropics, making them more susceptible to storm damage. Global warming and ocean acidification from increased CO_2 levels seem to be working in a destructively multiplicative manner on some reefs [22].

Coral bleaching happens when a spike in water temperature causes the cells of the coral polyps to expel tiny one-celled algae that live inside them. The algae capture the energy of sunlight and the coral polyps provide the algae with a safe place to live.

There has been an increase in mass bleaching events around the world since the middle of the twentieth century. If the corals are not damaged before the temperature drops again, they can recover by capturing new populations of algae. But if they are unable to lure new algal symbionts they will die and be smothered by different species of free-living algae. Reefs that had once been thriving can quickly become slimy green graveyards. The hard scleractinian corals, which form the backbones of reef structure, are especially susceptible to such damage [23].

Even under ideal conditions, it may take a decade for reefs to recover from a severe bleaching event [24]. As ocean temperatures rise, bleaching will inevitably happen more often, which reduces the chance of recovery before the next event.

Coral bleaching is a serious problem, but in the near future its effects are likely to be spotty and unpredictable. The greatest long-term threat to coral reefs around the world comes from increases in the amount of CO_2 that is dissolving into the oceans, where it combines with water to become carbonic acid.

Even when CO_2 molecules are dissolved in water, most of them remain intact. But a small fraction of them ionize and combine with water molecules to produce the bicarbonate ions of carbonic acid, HCO_3^-. A small amount of the bicarbonate in turn becomes further ionized to carbonate ions, $CO_3^=$. The hydrogen ions that are released during these ionizations increase the water's acidity.

At first glance, one might suppose that when CO_2 in the atmosphere increases, so that more of this gas dissolves in the world's oceans, this should increase the carbonate ions in the seawater. And this might seem to be a good thing for the creatures of the reefs. Corals and many of the other animals on the reefs can combine doubly-ionized carbonate ions with ionized calcium. The resulting calcium carbonate then crystallizes, and the crystals form the corals' hard skeletons. These tiny precipitation reactions, carried out by astronomical numbers of coral polyps and other animals over millions of years, have built great limestone structures such as the Great Barrier Reef, which stretches for over 2,000 kilometers [25].

But the laws of chemistry lead inevitably to a different result. It is these laws that dictate the relative concentrations of the carbonate and bicarbonate ions in the water, and they operate without regard to the needs of the organisms that are living in the ocean.

These chemical laws dictate that when the level of dissolved CO_2 increases, all those extra hydrogen ions in the water shift its pH and make it more acid. The chemical equilibrium between bicarbonate and carbonate ions shifts in response. And, alas, the shift is away from useful carbonate and towards bicarbonate, which the animals cannot use. The chemical building blocks that the reef animals need for healthy growth become progressively more unavailable as CO_2 builds up and pH drops.

Experiments and observations in the field show clearly that when the water's acidity increases and corals are stressed, they respond negatively. Massive *Porites* corals in the Great Barrier Reef have slowed their skeleton production by fifteen percent since 1990, a decrease that tracks the decrease in pH of the ocean's water [26]. In areas off the coast of New Guinea, some settled corals are exposed to unusually low pH because of nearby volcanic activity. The growth of most of the coral species in these areas has slowed dramatically, which allows a few resistant species to take over and lowers the reefs' overall diversity [27]. And laboratory studies have shown that the growth rate of the hard coral *Stylophora pistillata* is cut in half when CO_2 is doubled and when the water temperature is increased by three degrees centigrade. Such conditions are likely to be common in the oceans at the end of this century [28].

Coral polyps are amazingly tough—even after they have been stripped of their skeletons in the laboratory they can survive long enough to build their skeletons again [29]. But they are not infinitely adaptable.

Are rising temperature and falling pH enough to cause the deaths of the majority of coral species, and of the many other organisms on which the life of the oceans and of the whole planet depend? We do not know, but it is a sobering fact that such widespread mortality has happened several times in the history of our planet.

Many corals, including the entire ancient lineages of tabulate and rugose corals, died out during the greatest mass extinction during the history of multicellular life, the Permo-Triassic event of 245 million years ago. Corals were decimated again during the Cretaceous Period, from 145 to 65 million years ago, when dinosaurs were at the top of land food chains and large reptilian predators dominated the oceans.

"Cretaceous" means "chalky." The great Cretaceous chalk cliffs of Dover on England's south coast have been carved by the tidal races of the English Channel from a thick deposit of chalk that covers much of northern France and southern England. And this chalk in turn is made up of the calcium carbonate skeletons of tiny one-celled creatures, chiefly animal foraminifera and plant coccolithophores, which thrive in huge numbers in the warm oceans.

Warm is the operative word here. During much of the Cretaceous and the early part of the Age of Mammals that followed, the entire Earth was warm from pole to pole. There were no polar ice caps. For much of this time atmospheric CO_2 levels were up to three times as high as those of today [30]. (We will see what happened to all this CO_2 in the next chapter!) This was a world very like the world that awaits us as present-day CO_2 levels continue to rise.

One result of those high levels of CO_2 was a huge population explosion of foraminifera and coccolithophores. When they died their skeletons sifted down from the sunlit upper ocean into the depths and eventually built up those great chalk deposits. At the same time as the coccolithophores exploded in numbers, hard corals went into decline [30]. A similar event happened during the early part of the Age of Mammals, 56 million years ago, when there was another burst of warming and increased CO_2 [30].

It turns out that today's coccolithophores thrive at high CO_2 concentrations, producing thicker than normal skeletons. But when hard corals are grown at high CO_2 and temperature, as we saw, they lose half of their skeleton-building ability.

Why this difference? The coccolithophores form calcium carbonate crystals inside their cells, and later excrete them to construct their skeletons [31]. Corals, unlike the coccolithophores, must make the crystals for their skeletons outside their cells. As the coccolithophores build their skeletons, they have far greater chemical control over the process.

Some corals can use a variety of tricks, such as increasing the alkalinity in layers of mucus just outside their cells, to build up high concentrations of calcium carbonate [32]. It is these abilities that explain the resistance of some coral species to pH changes. It is possible that these more resistant coral species will be able to adapt to the present-day swift changes in the oceans, but any such evolutionary change would lag far behind the rate of ocean acidification. And such changes would inevitably involve huge coral mortality, as poorly adapted polyps die and are replaced by the descendants of a small minority of better-adapted ones. During this dangerous die-off period, the reefs might be smothered by explosions of better-adapted organisms such as the foraminifera and coccolithophores.

Reefs are not merely gorgeous places for tourists to frolic in. They are central to the integrity of the Earth's ecosystems. Without them, entire islands would be washed away. Continental margins such as the eastern coast of Australia would be repeatedly devastated by storm damage.

My geochemist colleague Dick Norris of the Scripps Institution of Oceanography has emphasized to me that we have no evidence for a direct connection between CO_2 levels and the waves of coral extinctions during the Cretaceous and the early Age of Mammals. But he considers it likely that warming and rising CO_2 levels played a role. He also points out that there is no precedent anywhere in the fossil record for a rise in CO_2 that matches today's rapid increase.

The world's reefs will only be protected from the existential threats that face them when we as a species face the consequences of our actions that change the very chemistry and physics of our planet. It may take a widespread and dramatic disaster that unequivocally demonstrates the existential threat of global warming before we wake up to our responsibilities.

In this chapter we have seen the challenges that face even small-scale efforts to restore green equilibria. Existential threats and other large environmental changes pose much bigger problems. Now, in order to lay the groundwork for understanding what might be involved as we try to restore the balance of entire ecosystems, let us look at how three very different ecosystems responded to such huge changes before the appearance of humans. The history of these challenges can give us valuable insights into how we can save the world of tomorrow.

··

CHAPTER FIVE

Catastrophes of the Past

How Three Different Ecosystems Have Responded to Environmental Challenges

The existential crises that currently face coral reefs and other ecosystems have the potential to transform our planet within a single human lifetime and make it far less friendly to life. In this chapter I will trace the ancient histories of three of the ecosystems that we are visiting in this book. Each ecosystem underwent transformative changes before the appearance of humans. Some of these changes happened suddenly, others unfolded over tens of millions of years. In each case, the ecosystems were able to recover and adapt, green equilibria re-emerged, and the world eventually became a richer place. But in each case the recoveries played out over millions of years.

We can draw three conclusions from these histories. The first is that the ability of the living world to adapt to severe crises is astonishing. The second is that the most profound of these adaptations take long periods of time. And the third is that we can draw lessons from the past about the future of our own species.

If our behavior triggers ecological disasters, a wonderful new world may indeed emerge in the long run. But, as John Maynard Keynes famously said, in the long run we are all dead. And if our species goes extinct as a result of our missteps, then all our progeny will also be dead before the planet has a chance to recover. The threat posed by severe existential crises makes it all the more essential that we get our act together. We must use the abilities that we have gained through our own long evolutionary history in order to avoid these crises in the first place.

The first of our three ecosystems is one that we have already visited. The ancient Guiana Shield is one of the few places on Earth with sufficient geological stability for some of the world's oldest lineages of animals and plants to survive. Even so, what we see today on the Shield is a palimpsest: a few fragments of surviving ancient life that are hidden beneath successive layers of more recent immigrants. The influence of humans, the most recent layer, consists of an almost microscopically thin film of time that spans a few thousand years. Our history overlies billions of years of earlier change.

Stability and Isolation

The country of Guyana, once part of the British Empire and now independent, is two-thirds the size of Italy—but it is home to eight hundred times fewer people. It occupies part of the Guiana Shield, one of the oldest pieces of continuously dry land on the planet.

The Guiana Shield stretches from Colombia to northern Brazil. It is one of a dozen or so *cratons* that are scattered across the planet. These ancient pieces of crust and upper mantle are descended from some of the first substantial bits of land that emerged from the worldwide ocean four billion years ago.

Other long-lived cratons include Greenland, the Canadian Shield around Hudson's Bay, and some pieces of southern Africa, Australia, and India. Cratons make up only about ten percent of the present-day continents. The remainder of the Earth's continental crust has been formed much more recently, by geological forces such as volcanic activity and the upwelling of mantle in the deep oceans.

The Guiana Shield craton is a huge mass of crust and solid mantle that pushes 200 kilometers down into the *asthenosphere*, the hot, liquid part of the mantle. Like the other cratons, it is twice as thick as more recently-formed pieces of the Earths' crust. This thickness and stability has helped it, an island in a storm of geological change, to survive the successive formation and breakup of the supercontinents Nuna, Rodinia, and Pangaea over the last two billion years.

Pangaea, the most recent of these vast supercontinents, began to split across its middle from west to east about 180 million years ago. The northern part, Laurasia, would eventually become North America, Europe, and Asia. The southern part, which included the ancient Guiana Shield, had already played a major role in the earlier formation of Pangaea. This southern region is known as Gondwana.

During all these events thick layers of sandstone, shale, and other sedimentary rock repeatedly built up on the Guiana Shield's ancient base and then eroded away. The latest round of this continual buildup and erosion left behind the *tepuis*,[1] one of which is now the home of the golden poison arrow frogs.

During one two-day period I encountered three species of animal that together span the most recent four hundred million years of the Guiana Shield's eventful history. Their stories show how the living world of the Shield has changed, adapted, and continually coevolved in spite of environmental catastrophes and the arrival of thousands of later invaders.

The first of these animals is the air-breathing arapaima, one of the world's largest freshwater fish. It can reach a length of three meters and a weight of 250 kilograms. And it is one of many fish species that require periodic shots of oxygen-rich air.

Young arapaima use their gills to extract oxygen from the water. But when they reach only a few millimeters in length their gills begin to lose this ability. The blood vessels of their mouth and throat enlarge to form a kind of lung. The maturing arapaima begins to swim to the surface periodically, gulping air and extracting oxygen from it.

Arapaima have been decimated by fishermen in Guayana's rivers and throughout the Amazon basin. But substantial populations of the fish still survive in some of the more isolated oxbow lakes. Liz and I, accompanied by our guide Rovin Alvin, landed on a sandbar a few kilometers up the Rewa, a major river that flows into the Rupununi. From there we hiked and waded two kilometers inland through swampy forest. At the end of the trail we found a mirror-calm oxbow called Grass Lake that supports the largest known population of arapaima anywhere in the Amazon region.

Arapaima are a challenge to photograph, because they spend so little time on the surface. They suddenly breach the calm water, gulp air, and vanish. No matter which way I pointed my camera, they perversely surfaced somewhere else. By the time I swiveled and focused on them, they were gone. It was like playing a giant game of whack-a-mole. Finally, after two days, through sheer dumb luck I caught an arapaima that was in focus and handsomely gaping (Figure 45).

The ancestors of most of today's bony fish had lungs, simple outpocketings of the throat that were heavily supplied with blood vessels. Now, most bony fish have converted those lungs into swim bladders that are no longer connected to the world outside [1]. These fish have gone back to the old gills-only system of breathing, but they have gained the ability to cruise effortlessly in their aquatic environment.

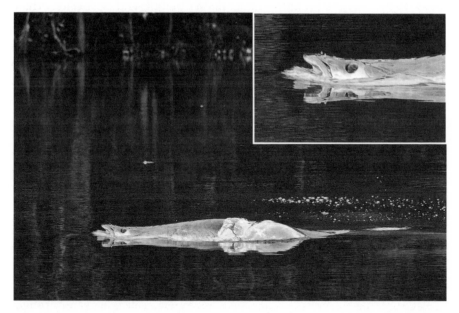

Figure 45 This arapaima fish surfaced briefly in one of Guyana's oxbow lakes, gulped air, and disappeared again within about two seconds. It was moving so quickly that the water formed a sheet over the middle of its body. The closeup at upper right shows details of the head of this primitive fish. Such living fossils have survived past catastrophes because of the remarkable stability of the geology of the Guiana Shield.

The arapaima has a normal fish-like swim bladder, but its mouth and throat with their numerous blood vessels have evolved into a primitive lung to supplement its gills. Although the details are different, the arapaima is recapitulating the much earlier evolution of lungs that had first taken place in ancient bony fish [2].

The arapaima that I had finally managed to photograph was demonstrating some of the ancient biological imperatives that had led to air-breathing. And that ability, which its ancestors gained long ago, enabled a few of them to survive a truly existential threat.

On what must have begun as a perfectly ordinary day almost sixty-five and a half million years ago, a meteorite with a mass of 1.3 trillion tons hit shallow ocean near what is now the Yucatán Peninsula. The impact threw up a vast plume of molten material that fanned out for thousands of kilometers to the northwest. The resulting tidal waves towered three hundred meters when they reached the northern edge of the Guiana Shield, two thousand kilometers from the point of impact. It was only the Shield's immense size that prevented it from being totally flooded. And, because this craton's ancient geological roots of solid rock penetrated deep into the Earth's molten asthenosphere, it was not torn apart by the thousands of volcanic eruptions that the impact triggered in other parts of the planet.

In the Shield's stable center, some ancestors of the arapaima, along with lung-fishes and a few frogs, toads, lizards, crocodiles, and turtles, burrowed into the mud to escape the impact's blast-furnace heat. The air that they depended on cooled as it penetrated into the mud.

When these lucky amphibian and reptilian survivors emerged, they found themselves in a world in which the dinosaurs, most of the mammals and birds, most other species of animals, many plants, and probably many members of the invisible world had been wiped out by the impact [3].

Although most mammal species had disappeared, representatives of at least twenty major groups survived [4]. Members of twenty-two of the major groups of birds also managed to live through the catastrophe [5]. Even as the survivors radiated adaptively into the millions of newly-emptied ecological niches, the arapaima and other ancient lineages were able to hang on to a few of these niches.

During our first afternoon of trying to photograph the elusive arapaima, we had made about a hundred attempts and failed at all of them. But in compensation for our failure, as we came back in the dark we stumbled on two other animals that helped to fill in a great swathe of the Guiana Shield's history.

The first of our accidental discoveries reinforced a profound rule of ecology and evolution that gives us reason for hope, even as we contemplate the worst devastation that our own species has inflicted on the living world. The process of evolution is so powerful that, even after the darkest moments in the history of life, it has helped the living world to rebound in new and unforeseen ways.

In the black depths of the forest we found a long-tailed mouse opossum. The tiny animal regarded us pensively from a branch, like a Cheshire Cat without the grin (Figure 46). This marsupial is distantly related to kangaroos, but it does not carry its young in a pouch. Immature baby opossums use mighty Popeye-like forelimbs to crawl straight from their mother's womb to one of her teats, where they hang on for dear life as they continue developing [6].

As the dust settled and the skies cleared, the surviving South American animal and plant species were truly alone. The continent had already been an island for tens of millions of years. It had drifted so far from North America and Africa that none of the mammals that survived on those equally devastated continents

Figure 46 This marsupial mouse opossum, *Micoureus demerarae*, is a survivor of one of South America's most ancient mammalian lineages. It stares at us from a branch above the trail leading from Grass Lake.

could cross the gaps. South America's isolation would continue for another fifty million years, allowing evolution to proceed in what paleontologist George Gaylord Simpson memorably called "splendid isolation" [7].

Released from the tyranny of the dinosaurs, the mammals and other animals that survived began to radiate adaptively into unique forms. They, along with the surviving plants, pathogens, and parasites, began to form complex ecological relationships that would eventually generate new green equilibria.

Marsupials had originally arrived in South America from North America more than fifty million years before the impact of the meteorite, and luckily they were a diverse crew. After the impact this inherited diversity helped them to evolve into a wide range of animals that could take advantage of the continent's newly empty ecological niches. The result ranged from small insectivores such as our little mouse opossum, through hyena-like borhyaenids with bone-crushing jaws, to big saber-tooth cats that from all the evidence looked exactly like their placental equivalents in North America.

The South American placental mammals that survived the meteorite also diversified. Unlike the marsupials, some of them took advantage of grazing and browsing opportunities that opened up with the spread of grasslands. Massive horse-like litopterns with elephant-like trunks raced across the newly opened country. The middle toes of some of these litopterns evolved into hoofs, in parallel with those of some horse lineages in Eurasia and North America.

Litopterns shared the plains with a variety of other large placental browsers and grazers, such as the immense *Toxodon*. When Darwin went ashore from the *Beagle* to explore Patagonia, he encountered recent *Toxodon* fossils and noted with astonishment that they seemed to be a blend of elephant, manatee, and giant rodent [8].

The grave demeanor of the little opossum on its branch was, we concluded, appropriate to the weight of its hundred and twenty million years of eventful history and its narrow escapes from existential threats.

We crept past the opossum and left it in darkness. Then, just a few yards further down the trail, we found another small mammal on a branch. This was a spiny tree rat (Figure 47).

Figure 47 Only a few yards from where we saw the marsupial opossum, a placental spiny tree rat (*Proechimys cuvieri*) crouches alertly. You can see the slender but sharp spines projecting through the fur on its back. These tree rats are relatives of the guinea pigs (which despite their name are also South American and are neither from Guinea nor pigs!) The ancestors of these placental mammals were among the first to end South America's splendid isolation.

About 30–40 million years ago a trickle of new placental mammals, including the ancestors of the tree rat, heralded the end of South America's long isolation. Some of the earliest of these immigrants apparently came from Africa, which was much closer to South America than it is now.

These *arrivistes* included lineages that would become the guinea pigs, capybaras, New World monkeys, and sloths. One branch would evolve into the awe-inspiring giant ground sloth *Megatherium*, a browser that reached six meters in length and weighed more than a bull elephant.

The first of these invaders may have helped to drive a few of the original South American mammals to extinction, but most of the indigenous mammals, including the litopterns, toxodons, and marsupial saber tooths, continued to thrive. Their reprieve, however, was temporary. When the Isthmus of Panama began to form, starting six million years ago, a trickle of island-hopping immigrants from North America was replaced by a flood.

These migrations, collectively known as the Great American Biotic Interchange, did not involve all the inhabitants of the two continents. The deserts of the North American Southwest and of Mexico formed a barrier that prevented a similar exchange among plants [9, 10]. And in South America the mammals that stayed behind managed to coexist with the flood of new northern animals for hundreds of thousands of years. But eventually the litopterns and the marsupial saber tooths faded away, along with many other long-time inhabitants of the continent.

Everywhere we went in the Guyanan forests and grasslands, we saw animals that had arrived during the Great Interchange.

Tapirs are amphibious placental browsers that once were widespread in Eurasia and North America. When the Isthmus of Panama formed, these amphibious animals were able to swim across its swamps and brave its jungles on their way to South America.

Today, tapirs survive in only two places: the rainforests of Central and South America and those of Southeast Asia. Even though tapirs are currently the largest native mammals in South America, they are seldom seen in the wild. So we were delighted when one found us.

We were photographing birds along the banks of Mapari Creek when our guide Ashley Holland recognized a tapir's mellifluous fluting calls. The calls came closer, and suddenly the tapir emerged from the forest, scrambled down the bank, and swam across the river (Figure 48). It passed within two meters of our boat, continued to the opposite bank, and climbed up through a cloud of sulfur butterflies.

Figure 48 Three million years ago the ancestors of this Brazilian tapir, *Tapirus terrestris*, arrived from North America and helped to displace the native litopterns and other browsers.

This featured player in the Great American Interchange is still found in substantial numbers in the Guyanan rainforests. In 2008 Holland led a group of scientists, sponsored by London's Zoological Society, in a series of portages up a group of cataracts to the previously unvisited headwaters of the Rewa. Every 500 meters along the river there were signs of tapir crossings. Remoteness has preserved the green equilibrium of these forests.

The meteorite that killed the dinosaurs was not the only existential threat to Guyana's forests and grasslands. Their animals and plants have survived at least four other major mass extinction events that threatened their very existence. One of these, the mass extinction at the end of the Permian 245 million years ago, was so severe that it took ten million years for the planet to recover [11].

During our explorations of Guyana's forests we found repeated examples of life's resilience, along with sobering evidence of how close existential threats have come to wiping out all multicellular life. Recovery was often slow, in part because of South America's long isolation. But sometimes new ecosystems can emerge with surprising speed from a mix of older ones, suggesting that with care and foresight we too might be able to build new diverse worlds from fragments of the old.

The forests of New Guinea are as rich biologically as the Guiana Shield. But in contrast to the Shield this richness does not include parts of ancient ecosystems. Instead it results entirely from a recent collision of ecosystems that have had very different histories.

When Ancient Worlds Collide

New Guinea, the world's second largest island, originated from the geological equivalent of a multi-car pileup. Its history begins with the fragmentation of Gondwana. The tectonic plate that would become Australia split from Antarctica and moved east. About forty-five million years ago it began to veer north, undergoing a series of complex collisions with the Philippine plate, the Southeast Asian plate, and two different arcs of oceanic islands that had formed at the southern margin of the Pacific plate [12].

All these smashing and grinding collisions resulted in the formation of a chain of brand-new mountains along the Australian plate's northern margin (Figure 49). These collisions, along with their attendant volcanic activity, built up the jagged and dauntingly steep mountains that I had scrambled over when I left the village of Yawan.

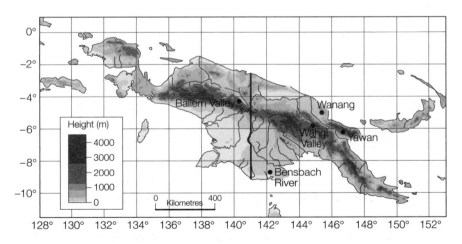

Figure 49 New Guinea's central mountain chain splits the island in two. The mountains mark a region of uplift where the Australasian plate to the south has collided with the Philippine and Pacific plates to the north. Labels indicate the villages of Yawan and Wanang, along with two of the isolated highland valleys and the southern Bensbach River.

These collisions brought together animals and plants that had traced out different evolutionary paths during the 180 million years since the breakup of Pangaea. The effects of the collisions are still visible. New Guinea's southern coastal lowlands reminded me, when I visited them in 2005, of a wetter and more tropical version of Australia. Mobs of wallabies hopped over the grasslands along the Bensbach River, and kookaburras called from the trees. In contrast, the lowlands along the northern coast of the island are home to many species of bird and plant that have island-hopped down from the Pacific island crescents. Plate 14 shows how three workers at the Wanang forest dynamics plot, Malape Kepe, Simbai, and Tony Yari, are dwarfed by the immense network of prop roots supporting a *Ficus archiboldiana* tree in this rich north coast lowland forest.

Because of the combination of all these ecosystems, it is not surprising that New Guinea is home to almost ten percent of the world's bird species, even though it makes up only 0.75 percent of the world's land area [13].

Unlike the ancient ecosystems of the Guiana Shield, the ecosystems of both the highlands and the northern slopes of New Guinea have become established on terrains that are rugged and recent. Because of the complexity of New Guinea's landscape, the number of available ecological niches is huge. As Vojtech Novotny's group has shown, their number is likely to be multiplied even further by host–pathogen and host–parasite interactions. But in some ways the world of New Guinea is gentler than the predator-rich forests of South America's Guiana Shield. Although fierce marsupial lions, possessors of the most powerful jaws of any known mammal, roamed Australia before the arrival of humans, there is no evidence as yet that they penetrated as far north as New Guinea's highlands or its northern slopes [14, 15]. It is true that a few fossils of small marsupial predators such as the cat-like thylacines have been found in the highlands, but in general New Guinea was not as terrifying a place as the broad grasslands and vast open forests of Australia to its south.

The relative paucity of predators in the highlands may have been a factor in the evolution of the birds of paradise. In these birds sexual selection for brilliant feathers and elaborate displays has been able to trump weaker selective pressures for camouflage and for more discreet behaviors (Figure 50).

To see how this happened, begin by considering the carefree displays of sexual selection that took place in a typical speakeasy during the era of prohibition. These displays could happen as long as the celebrants were confident that the owners had bought off the police.

Figure 50 At dawn, at the crest of the mountain that separates Yawan from the research station, I found a six-wired bird of paradise, *Parotia lawesi*, feeding on Araliaceae fruit. Birds of paradise have evolved their elaborate plumage and courting rituals during the last twenty-five million years, in part because they have few predators.

The spectacular colors and dances of the birds of paradise have evolved in just such a free-wheeling environment. DNA data show that the birds of paradise have been diverging in the highlands for at least the last 25 million years, suggesting that these relaxed, relatively predator-free conditions may have persisted for long periods [16]. I would like to suggest that one important reason for why we should do our utmost to preserve the green equilibria of New Guinea's forests is to ensure that the birds of paradise are free to party on!

In spite of the fact that New Guinea's ecosystems have a far shorter history than those of the Guiana Shield, the organisms that inhabit them have developed a robust green equilibrium based on thousands—and probably millions—of interlocking relationships. The abundant sunshine and rain of the tropics, the large size of New Guinea's lowland and highland habitats, and the fact that these habitats have been formed as a result of the collision of so many different biological provinces, have all aided the evolution of the island's rich species communities. Left undisturbed, these communities will certainly become even richer with time. And they show what we might accomplish in the future if we are forced to replace damaged ecosystems with new groupings of animals and plants, along with their invisible but essential world of symbionts and pathogens. Ecological purists may object to such an idea, but it is clear from New Guinea's history that nature has been there before us.

The Building of the Himalayas

Our third historical example is an ecosystem that has also emerged from a collision of tectonic plates. This new ecosystem has not only evolved in unique directions, but its emergence has had escalating consequences throughout the planet and has indirectly led to the emergence of our own species. The utter unpredictability of such events warns us that our own impact on the planet could take the future in equally unexpected directions.

The region involved is huge. It spans nothing less than the entire range of the Himalayas and the Tibetan plateau that lies to their north. The extreme altitudes of some of these regions challenge their inhabitants, including the humans who

Figure 51 I encountered this magnificent golden langur, *Trachypithecus geei*, in the warm subtropical foothills near Bhutan's southern border with India. These endangered animals have found protection in Bhutan from hunting, but their respite may be fleeting as Indian construction crews begin to widen highways and build hydroelectric plants in the Puna Tsang and Nangde river valleys.

Figure 52 At the other extreme, the tough and resourceful Himalayan Thar goats, *Hemitragus jemlahicus,* can graze during the summer in pastures above 5,000 meters. These goats, which survive in substantial numbers in Nepal's Sagarmatha National Park, are now the chief prey of the park's slowly recovering population of snow leopards [17].

live there, with strong evolutionary and ecological pressures. How did these ecosystems evolve?

Later in the book we will visit two countries, Nepal and Bhutan, that span these Himalayan ecosystems. The people who live in both countries have influenced, and been influenced by, these extreme ecosystems. But to understand the consequences of these interactions we must place these ecosystems in their evolutionary and ecological context. That context is a spellbinding tale in itself.

Within both Nepal and Bhutan it is possible to travel over the space of a hundred kilometers from subtropical lowlands to the fringes of the keen, cold boundary between our familiar atmosphere and the stratosphere (Figures 51 and 52). At this upper boundary, low temperatures and high altitude combine to define the outer edges of life's possibilities. It is remarkable that living organisms have been able to adapt to such extreme environmental conditions, especially when we realize that these conditions have rarely been present during Earth's long history.

In the autumn of 2009 I got my first close view of the highest of the Himalayas. Our plane flew to within about twenty kilometers of Mount Everest and

Figure 53 In this view of the Himalayas, the summit of Everest thrusts up behind Lhotse, the world's fourth highest mountain.

Lhotse (Figure 53). The two peaks, the highest and the fourth highest mountains in the world, are separated by only four kilometers.

Along our flight path Everest was partly shielded by Lhotse, whose great southern slope is perhaps the most difficult climb in the world, more challenging than Everest itself. It rises 3.2 kilometers over 2.5 kilometers horizontal distance, making it the steepest slope of its size anywhere on Earth. This huge slab of rock , like the rest of the range, was tilted upwards during the collision of two great *tectonic plates*.

Tectonic plates consist of a layer of crustal rock on top of a thicker layer of hot but relatively solid mantle material. As we saw, they float on the astheosphere, a deeper layer of molten mantle that lies beneath the crust everywhere on Earth. Two of these plates, one from Gondwana and the other a vast chunk of Laurasia, were involved in the collision that formed Lhotse and the other great Himalayan peaks.

The hot asthenosphere below the plates is in constant movement, like a pot of molasses at a rolling boil. Rising plumes of this molten rock can break up a tectonic plate, and such plumes were responsible for the breakup of Gondwana into several pieces.

These pieces of Gondwana would eventually become South America, Australia, Antarctica, Africa, Madagascar, and India. As Gondwana gradually broke up, the pieces were propelled in different directions by three major processes.

The first was the formation of new lithosphere in the thin crust of the ocean basins, where the molten asthenosphere could readily ooze to the surface. Geologists had first thought that this seafloor spreading was sufficient to push the continents apart. But tectonic plates are, to put it mildly, rather massive objects. Timid little pushes from thin new oceanic lithosphere as it forms and spreads are simply not strong enough to move them. It is now realized that two other processes, both involving huge convective cells and upwellings in the asthenosphere, play a more important role.

If two tectonic plates are riding different convective cells in the asthenosphere and meet head-on, the heavier one dives under the lighter one, pushing it up to make mountains such as the Himalayas and the jagged peaks of New Guinea. Such regions are called *subduction zones*. As the edge of the heavier plate plunges deep into the asthenosphere, it pulls the entire plate behind it. The process can be likened to a drunkard in a silent movie who falls from his chair and drags the tablecloth and all its settings with him.

This drunkard's pull of subducting plate edges helps to account for some of the movement of tectonic plates. But in order to produce a collision that was powerful enough to build the Himalayas and the Tibetan Plateau, something additional had to happen—in this case a massive push that originated deep in the molten asthenosphere.

About 160 million years ago, the whole eastern half of Gondwana began to split off from the western region that would become South America and Africa. This eastern piece, which would eventually split further and become Antarctica, Australia, Madagascar, and India, moved away at a speed that was about average for a tectonic plate, about four centimeters per year.[2] Thirty million years later the western piece also began to split slowly into Africa and South America.

All these movements were pretty normal for the world of tectonic plates. If a vast immortal intelligence had glanced at the Earth from outer space every few million years during this period, it would have seen nothing unusual.

Then, about 67 million years ago, a piece of the eastern plate broke away and sped up dramatically. This piece, which would eventually become India, moved swiftly north through what was then the Paleo–Tethys Ocean. It reached blistering velocities of up to eighteen centimeters a year, as far as we know the world speed record for tectonic plates. As India sped up, the African plate began to slow down.

My colleagues Steven Cande and David Stegman at the Scripps Institution of Oceanography have shown that India's speedup and Africa's slowdown were both caused by an upwelling plume of molten asthenosphere that originated beneath India's west coast. The plume gave India a huge push towards the north-east, at the same time as it slowed Africa's motion [18].

As the Indian plate moved swiftly north, its northern rim began to sink beneath the Asian plate. Then, as the plume's push diminished over the follow-ing ten million years, the drunkard's tablecloth effect began to predominate. India was pulled towards the mainland of Asia by the forces of subduction.

Our analogy of the drunkard's fall has its limitations, because the drunkard stops when he reaches the floor. But the Indian plate's edge kept descending far down through the gooey asthenosphere, pulling more and more of the plate after it. At the present time the lower, denser part of the Indian plate has plunged more—perhaps much more—than 200 kilometers down into the astheno-sphere. It is not alone. In other parts of the world pieces of old oceanic litho-sphere have been detected more than a thousand kilometers beneath the surface, still heading even deeper into the mantle.

The immense subduction process buckled the edges of both the Indian and the Asian plates into a huge buildup of crumpled and tilted rock. Seafloor depos-its that had been pushed ahead of the Indian plate folded over like layers of snow ahead of a snowplow, adding to the accumulating mass of material. In the north-ern part of the collision zone the Asian plate was levered up to form the immense Tibetan plateau.

It is not surprising that some of the region's legends may have been inspired by traces of these violent events. Valmik Thapar, in his superb book *Land of the Tiger*, recounts a story from the *Mahabharata*, one of the great Hindu epics [19]. It tells how Vishnu visited the northern shore of a wide sea (the Paleo–Tethys Ocean, perhaps?) and took pity on seagulls whose eggs were swept away by storms. In order to help the seagulls Vishnu swallowed the sea and revealed the Earth. Then, when Vishnu slept, a demon called Hiranyanksha attacked the Earth and made love to her so violently that her limbs—the Himalayas—were hurled into the air.

This legend, so unnervingly close and yet so delightfully far from geological reality, evokes erotic love to explain the Himalayas' stark and contorted geologi-cal strata. I wished after reading the tale that I could have brought the same poetic magic to the real story of the formation of these mountains, but I console myself with the thought that the real story is amazing enough.

Vast Collision, Vast Consequences

The first result of this great collision was the Tibetan Plateau, an uplifted area one quarter the size of the continental United States that has an average altitude of more than 5,000 meters. As the plateau's double-thickness lithosphere continued to build up, it resisted further deformation. Because of this resistance, starting about fifteen million years ago, material from the Indian plate began to push up even further along the southern and western boundaries of the plateau. The result was the Himalayas and the almost equally massive Karakoram Range of Pakistan.

These ranges include all fourteen of the world's mountains that are more than 8,000 meters tall, along with hundreds of peaks that reach to more than 7,000 meters. The force of the collision has diminished as the Indian plate has gradually slowed, but the Himalayas continue to grow at a rate of four centimeters each year.

Because these events are relatively recent, the new ecosystems that they created have not yet reached equilibrium. Relatively few birds and animals can survive at such extreme altitudes. In Mongolia, to the north of the plateau, I encountered two of them: the Lammergeier or bearded vulture, and the red-billed chough (Figure 54). The Lammergeier is a scavenger that is found throughout the entire region, and the chough, equally widespread, holds the songbird record for high-altitude flight. These little birds have been seen on Everest, flying past parties of climbers at almost 8,000 meters.

A report from a 1963 American scientific expedition to Everest was the first to describe the Tibetan plateau and its fringing mountains as the Earth's "Third Pole." The plateau harbors the largest collection of glaciers outside the actual polar regions. At least 45,000 distinct glaciers, with a total volume of about 5,600 cubic kilometers, are scattered across 100,000 square kilometers.

The tectonic collision between India and Asia not only made these glaciers possible, but may also have brought on the ice ages that have buried the Arctic and Antarctic under much larger masses of ice. We have to look back 250 million years, to a time when Pangaea was still intact, to find a comparable event. At that time a massive uplift in southern Africa led to extensive glaciations that covered what is now the Karoo [20].

Figure 54 The Lammergeier or bearded vulture (*Gypaetus barbatus*) is a wide-ranging scavenger that can reach high altitudes, and the red-billed chough (*Pyrrhocorax pyrrhocorax*) holds the song bird altitude record. I spotted the vulture in central Mongolia's Yol Valley, and the chough near Genghis Khan's ancient capital of Karakoram, but both species range widely across the Tibetan plateau and the Himalayas.

The Indian lithospheric plate began its long collision with Asia about 52 million years ago. For most of the previous 50 million years, the entire planet had been steamy and tropical from the equator to the poles. Then, at the start of the collision, the first glaciers appeared in Antarctica [21].

During the collision period the cooling trend accelerated. Its cause may simply have been a change in weather patterns, triggered by the uplift of the Tibetan plateau which prevented monsoon rains from falling in central Asia [22]. But some geologists suspect that the primary source of cooling may have been a change in the balance of chemical reactions in the Earth's crust [23].

As the great plates collided, they brought enormous regions of long-buried rock to the surface. These rocks were rich in calcium aluminum silicate, which reacted with CO_2 in the air and water to produce the insoluble molecules calcium carbonate and aluminum silicate. This vast withdrawal of CO_2 from the atmosphere may have resulted in a reverse greenhouse effect, pushing the entire planet towards lower temperatures.

Thus, to an astonishing extent, the world as we know it may have been shaped by the collision of India with Asia. In particular, the cumulative effects of the collision on world CO_2 levels may have helped to trigger the repeated ice ages that began about two and a half million years ago. The great fluctuations in climate that resulted spread far beyond the glaciers. In the next chapter we will see how they may have contributed to the origin of our own species in equatorial Africa.

In this chapter we have followed the histories of three of the world's great eco-systems. These ecosystems have survived vast changes, even existential crises, and have arisen renewed like the Phoenix of legend. The recovery period has sometimes been achingly slow. These histories serve as a warning to us—if we trigger existential crises, their consequences may extend far beyond the lifetime of our own species. But we have also seen that life is so resilient and adaptable that ecosystems can be blended together to form something altogether new, as has happened with the formation of New Guinea. This gives us hope that we can not only undo the damage we are doing but can perhaps create something new and wonderful in the process.

These surveys through deep time also provide context for the ways in which our own evolution has been shaped by geological and biological history. As we will see next, our history is intertwined with the histories of the rest of the living world.

CHAPTER SIX

A Blending of Genetic Equilibria
The Origins of Our Species

New Guinea's highlands are famous for their boutique coffees. A combination of carefully selected strains and a perfect coffee-raising climate produces especially flavorful beans. Like most writers I depend on coffee to stimulate my flagging intellect, but such exotic coffees are out of my price range. I rely on blends, with their tradeoffs of flavor and price.

Blending, with its compromises and tradeoffs, has also played an essential role in ecosystems and in gene pools. We saw in the last chapter that New Guinea's complex living world arose as a blend of the highly divergent Australian and Pacific ecosystems, with the added dimension of a recent round of mountain-building that has introduced numerous new ecological niches. These blended ecosystems permit the growth of wonderful coffees, and they also support some of the most diverse and fascinating human societies on the planet (Figure 55).

In this chapter we will see how, like a mixture of coffees, the gene pools of these peoples have mixed in ways that have opened up, and will continue to open up, new opportunities for our species. And in the chapter that follows we will see how such blending has given us the genetic potential to understand, maintain, and restore the balance of genetic and ecological equilibria in the world that we inhabit.

To make a blended coffee, you must start with diversity. The same applies to a blended gene pool. For most of our history as a species, different groups of humans lived for long periods in relative isolation. They evolved in different directions, driven by chance events and the diverse pressures of natural selection. And their genetic radiation has been accompanied by, and sometimes driven by, cultural radiations.

When I walked through the neat and flower-filled gardens of the highland village of Yawan and began the climb towards the Ubii Camp, I thought about

Figure 55 These Dedua tribespeople from Papua New Guinea's Huon Peninsula are celebrating at the Mount Hagen sing-sing. Their celebration includes traditions that span fifty thousand years. Ancient Melanesian cultural patterns have been overlaid by a more recent Pacific island heritage. At the same time as this cultural blending was taking place, genes from New Guinea's Melanesians and from far more ancient peoples were entering the gene pools of the Austronesians and Polynesians of the Pacific. In this chapter we will look at such gene flows, and show how they have contributed to the genetic equilibria that have helped our species adapt.

the way in which this remote and jagged landscape has produced a vast cultural experiment that in turn encourages genetic diversity.

Suppose, I imagined as I puffed up the slope, we could conduct an experiment in human cultural development that was the equivalent of the settlement of New Guinea. Suppose we could transport a collection of young children to a different planet, and settle them on a vast mountainous island surrounded by ocean. Suppose further that the climates of the island range from steamy humid tropics on its coasts to chilly, windswept high-altitude tundra near the mountain summits. And let us say that the island is bursting with life—over much of its area there is an abundance of roots, fruits, and easily caught birds and animals, so that the children will not starve.

Imagine that we teach these little colonists how to make only primitive tools: stone axes, knives, spears, and fish-nets. They will be allowed no books, or any of the other accoutrements of the information age. Then we climb into our spaceships and depart. What would happen, culturally and genetically, to the descendants of these children over the next few centuries and millennia? What would our descendants find if they were to visit the planet fifty thousand years later? The results of our experiment would depend on the genetic legacy that was carried by these involuntary colonists, and on the fifty thousand years of further diversification and cultural fragmentation that took place on the island.

Exactly such a vast social and evolutionary experiment (though without the spaceships) has been carried out on the island of New Guinea, along with a similar experiment on the nearby but far different continent of Australia. Except for New Britain and a scattering of other islands to their east, these are the furthest points that our ancestors were able to reach during the longest and most recent of their great migrations out of eastern Africa. This migration, which took place between 100,000 and 50,000 years ago, covered about 16,000 kilometers overland.

The story of this far-ranging migration, remarkable as it is, is only part of the story of how modern humans left Africa. We now realize that its course was influenced by a series of similar migrations of earlier peoples. I last traced what was known of those earlier migrations in 2010, in *The Darwinian Tourist*. Now, a mere three years later, we have learned far more about these earlier waves of peoples as they first ventured into a world so utterly unlike their African home. And we now have new and astonishing details about genetic exchanges that took place between the earlier migrants and the modern humans who followed them [1]. It is as if the spaceships that carried our little colonists, on their journey to that remote planet, had encountered other spaceships carrying different groups of colonists, and the groups had been able to fraternize.

To put these migrations into perspective, and to understand why they were so transformative, we must start at the beginning.

The Origin of Our Species

About seven million years ago, an evolutionary ferment began among some of Africa's ape lineages. This ferment coincided with, and may have been triggered by, an increase in the variability of the world's climate. Geological and fossil indicators show that temperature and rainfall throughout the world were becoming more unpredictable, perhaps in part because of the buildup of the Tibetan Plateau and the Himalayas [2]. Ecological niches appeared, disappeared, and shifted from place to place. These shifts may help to explain why a great variety of primates, presumably able to adapt to these changing environmental patterns, appeared in Africa.

Most of these primate lineages are now extinct. They have left behind a fossil record that is fragmentary but still quite rich and informative. Taxonomists have grouped these extinct African genera, together with our own genus *Homo*, into the subtribe Hominina, which is colloquially known as the *hominans* [3]. From oldest to youngest these genera are *Sahelanthropus*, *Orrorin*, *Ardipithecus*, *Australopithecus*, *Kenyapithecus*, and finally *Homo*.

The brains of all these hominans, except for *Homo*, were ape-sized. The brains of our ancestors only began to increase markedly in size with the emergence of *Homo habilis*, the earliest member of our genus. The size increase started about two and a half million years ago at the beginning of the recent series of ice ages. But this increase was only part, and a relatively recent part, of the story of hominan evolution. Hands like ours probably first appeared in some lineages of *Australopithecus*, and it is likely that Australopithecines were among the first to make crude stone-cutting tools. And many other human-like characteristics can be traced even further back in time. It is likely that all of the hominans that we know of were able to walk upright more easily than chimpanzees and gorillas, and human-like dentition also evolved early.

Some of the fossils that preserve an early part of this history are the remains of seventeen individuals of *Ardipithecus ramidus*, a small-brained hominan that lived about 4.4 million years ago in open gallery forests in what is now northern Ethiopia. Even at this early stage, *A. ramidus* was well on the way to a human-like set of capabilities and perhaps even a human-like social structure. These hominans' big toes were still strong and flexible, retaining the ability to grasp tree limbs, but this did not prevent them from walking upright. Their hands showed no sign of the rigid structure that is characteristic of the knuckle-walking hands of chimpanzees and gorillas. And they had already lost (if they ever had) the large and threatening upper canines that are maintained by sexual selection in male great apes. This loss suggests that *A. ramidus* lived in social groups in which male mating success was less dependent on overt threats than it is in the great apes [4].

There are three possibilities for how *Ardipithecus* evolved. The most traditional scenario is that there was a sudden burst of evolution that thrust them and the other early hominans away from a lineage of chimpanzee-like apes. The ancestors of chimpanzees then proceeded on their own little-changed evolutionary path towards present-day chimpanzees and bonobos.

A second possibility is that the human-chimpanzee ancestor already had some hominan characteristics. These were subsequently lost in the chimpanzee lineage, as it embarked on its own eventful evolutionary course towards chimpanzee-hood.

Third, perhaps the ancestors of humans and chimpanzees were different from both. In this scenario, perhaps the least likely, the hominan and chimpanzee lineages each underwent great changes as they diverged from their common ancestor. Many of the large changes in the hominan ancestry would therefore have taken place before the time of *Ardipithecus*.

We cannot decide among these possibilities, because almost nothing is known about the fossil record of chimpanzees. What is clear is that even the earliest hominans were undergoing great evolutionary changes.

Complicating the picture is a new story, emerging from studies of our DNA, about how the divergence between the human and chimpanzee lineages took place. The story casts light on the process of speciation itself, and shows how it can sometimes involve both the blending of gene pools and their subsequent sorting-out through natural selection.

Introgression, Speciation, and Genetic Equilibria

When my students and I first began to use the DNA sequencing method that had been pioneered by Fred Sanger in the 1970s, it was a messy process that involved dangerously high voltages, fragile sheets of gelatinous material, and a great deal of radioactivity. It took a day of hard work to obtain a few hundred bases of DNA sequence.

Since that time, the ability to sequence DNA has leapfrogged to the forefront of rapid technological change. Intel cofounder Gordon Moore predicted in 1965 that the number of transistors on a computer chip would double every two years. His prediction has come approximately true, and it has led to a total transformation in the way that human societies deal with information. But Moore's law pales into insignificance compared with recent advances in DNA sequencing technology. Since the introduction of next-generation sequencing machines in 2007, the cost of sequencing DNA has dropped by a factor of 30 every two years, fifteen times as fast as Moore's law. Intensive hunts for genes that contribute to cancer and mental illness, spanning the entire three billion bases of our genome, have now become feasible.

It is also now possible to extract DNA from old bones and find enough undamaged molecules to obtain substantial amounts of usable sequence. This capability has opened up a new window into genetic studies of our origins. Among other things, we can look for signs of *introgression*.

Introgression takes place when species exchange genes or acquire genes from other species, through hybridization or other genetic mechanisms. Such exchanges have been common throughout the great evolutionary tree of life. Some of the most startling examples of introgression are found in bacteria and in other bacteria-like organisms.

The Graeae were ancient crones of Greek legend. They passed around their last remaining eye and tooth so that each of them could take turns seeing and eating. Bacterial species behave rather like the Graeae, passing around viruses, transposable elements, small extra chromosomes, and even bits of alien DNA that the bacteria eat and occasionally incorporate into their chromosomes. These introgressions of genetic information among species of bacteria have contributed to the rapid emergence of new pathogenic strains and of strains that are resistant to many antibiotics.

In complicated multicellular organisms such as ourselves, introgression usually takes place through a more familiar mechanism: individuals of different species meet, mate, and have offspring. Genetic signals of such introgression have been found in the DNA of many plant and animal species. This raises a fascinating question. Has there been introgression in our own ancestry, and if so what were its effects? For example, could introgression between the ancestors of humans and chimpanzees have contributed to the evolution of human-like traits in *Ardipithecus*? Could there have been later episodes of introgression among the different groups of hominans that left Africa and began to fan out across the world? And might these introgressions have contributed to the unparalleled success of our species?

Evidence is growing that introgression has indeed played a role in our evolutionary history, starting with the events that led to the separation of the ancestors of humans and chimpanzees. In 2005 Nick Patterson and his colleagues at M.I.T.'s Broad Institute began to compare the human and chimpanzee genomes, which had recently been sequenced. Their goal was to follow the steps by which these two species had separated some six or seven million years ago [5]. Had the separation been sudden and clean, or had there been some genetic exchange— some introgression—between the two nascent species before they diverged so far from each other that exchange was no longer possible?

Speciation is a tricky concept. Darwin made the role of evolution in the formation of species the centerpiece of his *Origin of Species*. But he was unable to propose a mechanism to explain why, when two species arose, they often became clearly separated from each other. He tentatively suggested a "principle of divergence," through which species have evolved to occupy distinct parts of their environment. But he was not able to close the argument, because although he knew that there were barriers to mating between species he did not know how they arose.

In 1865, soon after the publication of Darwin's *Origin*, the Moravian monk Gregor Mendel published a paper setting out the rules of genetics. His discovery was ignored by the world for a third of a century, but his genetic rules would eventually lead to an understanding of how the divergences between species arise and are maintained. At the same time, our growing understanding of the nature of ecological niches has provided an ecological dimension to the process by which species are selected to become different from each other.

One kind of speciation, which Darwin had observed on the Galápagos Islands, is relatively easy to understand. The thirteen or more finch species that currently live on the islands can be traced back to a few common ancestors. Those ancestors were members of a finch species that had accidentally been carried to that isolated archipelago from the South American mainland about 1.6 million years ago [6]. Because these migrants were now completely isolated from the mainland, the allele frequencies in their little gene pool began to shift through mutation and through selection for adaptation to the conditions that they encountered on the islands. Gradually the descendants of the original immigrants evolved into new species. Eventually they changed so much that if some of them had been transported back to the mainland, they would have been unable and/or unwilling to mate with the mainland finches.

This separation of Darwin's finches from their ancestors on the mainland, and their subsequent divergence, is a clearcut example of *allopatric* ("other country") speciation. Allopatric speciation takes place as a consequence of geographic isolation, which prevents introgression between the diverging species.

Most speciation, however, is much messier. Allopatry does not explain why Darwin's finches continued to speciate on the islands where they landed, even though the birds on an island were not geographically separated from each other. And most new species do not arise on isolated islands, but in complex and vast ecosystems such as coral reefs, forests, wetlands, and savannahs.

In these ecosystems, if a new ecological opportunity opens up right next door to where a species lives, some members of that species may be better able to take advantage of it than others. But, because the other members of the same species are still nearby, they are still able to mate with them. When the members of a species can interbreed freely, then their gene pools will stay mixed together and speciation will not take place.

But suppose there is some kind of barrier between the groups, so that genes cannot flow with perfect freedom between them. This might permit the groups to begin to diverge genetically, as they adapt to the different niches.

Unlike the clear geographic separation that we saw with Darwin's finches, the barrier might not be strong enough by itself to allow the process of speciation to begin. But it might be strong enough to encourage the evolution of *isolating mechanisms* that could "kick-start" the speciation process.

Isolating mechanisms can be of two types, depending on when they act. They may act on the potential parents, making them less likely to mate and produce hybrid offspring. These are *prezygotic isolating mechanisms*, because they act before the hybrid zygote forms from the union of sperm and egg cells. Or they may act on the hybrids themselves, lowering their ability to survive or reproduce. Such mechanisms, which exert their influence after the parents have incautiously mated, are *postzygotic isolating mechanisms*.

Suppose that hybrids that arise from matings between these two nascent groups are at some kind of disadvantage. This may be because they have a reduced ability to utilize either the old or the new ecological niches effectively, or because they suffer from genetic incompatibilities that reduce their fitness or fertility. Such postzygotic isolating mechanisms in the hybrids cannot themselves be selected for. This is because their harmful effects only appear in the hybrids after their parents have made what turns out to be a mistake by mating with each other.

By the time the parents have mated, it is too late to undo the damage. But if there is postzygotic selection against their hybrid offspring, it immediately generates a strong selective pressure for the accumulation of prezygotic mechanisms. These mechanisms are selected for because they encourage the members of the two groups to mate preferentially with other members of the same group. If they do, they won't waste their genes producing ill-adapted, sick, or sterile hybrids.

Sometimes it is possible to catch what may be the beginnings of a speciation process that involves these isolating mechanisms. Liz and I stumbled on one such case as we drove through the savannahs of Namibia in southwest Africa in 2008. Social weaver birds, *Philetairus socius,* build massive colonial nests in the acacia trees that dot the landscape (Figure 56). The nests last for years and provide a safe refuge for huge colonies of birds. But eventually the nests become so heavy that they crash to the ground, sometimes bringing down entire trees.

Figure 56　This Namibian social weaver bird nest can provide a home for other bird species, adding ecological niches to their environment and possibly stimulating speciation.

When we examined some of the nests, we discovered that the weaver birds have become accidental landlords of peach-faced lovebirds, *Agapornis roseicollis* (Plate 15). We learned later that our observation was not new. Others had noted that while most peach-faced lovebirds build their own nests, a minority occupy and modify nests within weaver bird colonies [7].

The nest-building behaviors of African lovebirds vary from species to species, and are influenced by genetic differences among the species [8]. It has been known for a long time that hybrids between various African lovebird species have difficulties with nest building, because they are unable to use the nest-building techniques of either of their parents [9]. Thus, the two groups of Namibian peach-faced lovebirds, those that build their own nests and those that occupy the weaver bird nests, might have accumulated genetic differences that have led to their behavioral differences.

One of these groups continues to be successful at building their own nests, while members of the other group have been selected for a tendency to move into pre-owned weaver bird nests. Perhaps hybrids between them might be less successful at either kind of behavior.

One way to test this possibility would be to measure the nest-building behavior of hybrids between the two groups. If their behavior does not fit into either pattern, thus harming their ability to reproduce, this would be a postzygotic isolating mechanism that separates the two groups of lovebirds.

It is also possible to ask whether members of the two groups mate freely with each other, or whether they tend to avoid each other. Avoidance would indicate that they have accumulated prezygotic behavioral mechanisms that prevent mating. Many different kinds of prezygotic mechanisms can be selected for, including differences in courting behaviors, differences in appearance, and differences in biochemistry that might influence mating pheromones. Sexual selection often plays a powerful role in the accumulation of prezygotic mechanisms [10].

Although there is as yet no evidence that these Namibian lovebirds could be splitting into two species, the potential for speciation is there. If the birds sort out into two groups, one that prefers to nest by themselves and another that seizes the opportunity to become squatters in the weaver bird nests, these behavioral differences can become reinforced over time as prezygotic mechanisms are selected for and postzygotic mechanisms continue to accumulate [11].

This type of speciation is called *sympatric* or *parapatric* speciation ("same" or "next" country). The distinction between the two terms depends on the degree of physical separation between the two nascent species. Like allopatric speciation, these types of speciation depend on a barrier of some kind that divides organisms into two groups. But sympatric or parapatric speciation can take place even if the genetic barrier is a porous one, so that some genes can still flow between the groups.

If genes flow too freely between the groups, or if the physical differences between the adjacent niches are not great enough, speciation will not take place. But under the right genetic and ecological conditions a single gene pool can split into two.

Genes that are responsible for different kinds of pre- and postzygotic mechanisms have now been identified in many plant and animal species. These studies show that there are many genetic paths to sympatric and parapatric speciation [12–14]. Even if species arise allopatrically, they can accumulate prezygotic isolating mechanisms, though the accumulation takes longer because it happens by chance and is not driven by postzygotic mechanisms [15].

It is clear that humans and chimpanzees are different species at the present time. We are separated from chimpanzees by strong pre- and postzygotic isolating mechanisms. But how did this speciation begin? Was it allopatric, parapatric, or sympatric?

Patterson and his coworkers found that some regions of the human and chimpanzee genomes were much more similar to each other than they should have been after six million years of separation. Was this evidence for a slow partitioning of the ancestral gene pool, such as would have happened during sympatric or parapatric speciation? The data were suggestive, but a clean allopatric split could not be ruled out [16, 17].

One feature of the data, however, did strongly support the possibility that our separation from chimpanzees was sympatric or parapatric rather than allopatric.

The amount of divergence between either human or chimpanzee X chromosomes and the X chromosomes of gorillas or orangutans is about what would be expected from the overall amount of DNA divergence between the species [18]. But when human and chimpanzee X chromosomes are compared they turn out to be much more similar to each other than they should be. They have diverged only about eighty percent as much as other matched pairs of human and chimpanzee chromosomes. This suggested that human-chimpanzee speciation was somehow different from the splits, millions of years earlier, that had given rise to the gorillas and the orangutans.

The sexes of humans and most other mammals are determined by the combination of X and Y chromosomes that they carry. Females of both species carry two X chromosomes, and males carry an X and a Y.

The X and Y pair, unlike our other pairs of chromosomes, are unusual because they are so different from each other. Males and females have two copies of all the genes on most of their chromosomes, but XX females have two copies of each of the genes on the X while XY males have only one copy. This is because Y chromosomes carry few genes, and most of them are different from genes on the X.

Patterson and his colleagues suggested that the low divergence between the human and chimpanzee X's could be explained if there had indeed been early hybridizations between members of the two nascent human and chimpanzee lineages. The hybridizations would still have been possible, even though prezygotic and postzygotic isolating mechanisms were beginning to accumulate, if isolation between the lineages was not yet complete.

The effects of hybridization would have depended on the genes that were beginning to accumulate in the two nascent gene pools. Suppose that a mutant allele appeared and spread on the X chromosomes of the group that will one day give rise to humans. The allele still functioned, and was not harmful in its own genetic context even when it was in males and present only as a single copy.

At the same time, an allele of a gene on a different chromosome appeared and spread in the chimpanzee ancestors. It too had no harmful effect on the individuals that carried it.

Now suppose further that these different alleles carried by the two groups, which had never met since they arose, would have interacted badly with each other if they did come together in hybrids. Figure 57 shows what would have happened when the two genes were brought together for the first time.

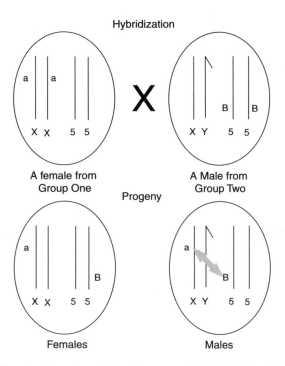

Figure 57 This diagram shows how X chromosomes might get lost in hybrid populations. When the ancestors of humans and chimpanzees began to diverge, the proto-human group (left) might have acquired an allele **a** on its X chromosome that has no harmful effect in its own genetic context. The proto-chimpanzees (right) might have acquired an allele **B** on chromosome 5. In its own genetic context the **B** allele also has no harmful effect. In hybrids, these two genes meet for the first time. In the hybrid females (left), there is no effect of **B** on **a**. This is because the **a** allele from the proto-humans happens to be *recessive* in its new context and is masked by the other allele on the X from the proto-chimpanzees. But in hybrid males there is a harmful interaction between the two genes (gray arrow), because the single copy of the **a** allele is not masked by another allele. If these damaged males die or fail to reproduce, their **a**-carrying X chromosomes will disappear from the hybrid population. In subsequent generations, the **a**-carrying X's might be lost entirely through a combination of selection and chance.

Because males have only one X chromosome, the allele on the X would interact badly with the new allele from the proto-chimpanzees in male hybrids and the hybrids would suffer. In subsequent generations, the proto-human X might have been lost, especially if the population size was small.

If such episodes happened repeatedly, they would have slowed the rate at which X chromosomes diverged in the populations that would eventually end up as humans and chimpanzees. Patterson and his colleagues estimated that there might have been a period of a million years, following the start of speciation, during which there could have been such genetic exchanges between these diverging groups.

This scenario for early human-chimpanzee divergence is hypothetical, but exactly such a situation has been found in other pairs of closely related species that have an X-Y sex determining system. When crosses are made between two such species, it is often found that the male hybrids die or are sterile, a phenomenon known as the Haldane effect. In some cases this effect can be traced to harmful interactions between allelic forms of genes on the X chromosome of one species and those on different chromosomes of the other species. These genes have never seen each other until they meet up in the hybrid [19].

The anomalously slow rate of divergence of our X chromosomes suggests that the parting of our gene pool from that of chimpanzees was messy and complicated. But there is no doubt that, if repeated introgressions took place, our gene pool would have been enriched. During that million years there would have been opportunities for genes on all the chromosomes of the two diverging groups to recombine with each other. Recall our discussion in Chapter Two about the amazing ability of genetic recombination to reshuffle alleles into new mixtures. Introgressions would have resulted in many new combinations of genes in our early gene pool.

Introgression may not have been confined to the time when humans and chimpanzees began to diverge. Recall that many closely related hominans—*Sahelanthropus, Ardipithecus, Australopithecus,* and *Homo*—began to appear in Africa between six and two million years ago. How much genetic exchange took place between these hominans as they speciated? And did these genetic resources contribute to the success of those brave hominan migrants who were the first to leave their home continent of Africa? As we will see in the next chapter, those early migrants needed all their genetic resources to survive and adapt to the new worlds that they were discovering. And we now have hard evidence that introgressions continued to happen to them even after they left Africa. All these events must have contributed to the complex genetic equilibria that have enabled our species to be so adaptable.

CHAPTER SEVEN

Ex Africa Semper Aliquid Novi

The *Homo* branch of the hominans first appeared about two and a half million years ago, in East Africa. These people were probably tool-users. They had brains that were substantially bigger than those of chimpanzees, and more than half as large as our own. The first of them, *Homo habilis*, soon gave rise to, or were replaced by, a new lineage with even larger brains, *Homo erectus*.

Some small bands of *Homo erectus* left Africa almost two million years ago. Their descendants traveled, over many generations, through the Middle East, central and southern continental Asia, China, and Southeast Asia (you can see the route that they followed in Figure 58). The descendants of these early *H. erectus* migrants are now all gone, but some of them survived on the Indonesian island of Java (marked with a 1 in Figure 58) down to as recently as 40,000 years ago [1]. And, as we will see, it is possible that through a devious route of introgressions these peoples have bequeathed some of their genes to some of us.

There are also slight hints of a simultaneous or even earlier migration out of Africa of people who were less like modern humans than *H. erectus*. The fine dotted line in Figure 58 traces this hypothetical early migration of small-brained Australopithecus-like hominans.

There are two pieces of indirect evidence for such a migration. First, a mysterious group of one-meter tall people, *Homo floresiensis*, (the "Hobbits") lived until recently on the island of Flores in Indonesia (marked with a 2 in the figure). The Hobbits had wrist and foot bones that resembled those of the small-brained African Australopithecines [2]. Their skull morphology, however, resembled that of *H. erectus* [3].

Second, some skeletal remains from western Asia, dating from 1.7 million years ago, also show a mysterious mix of affinities to both Australopithecines and later hominans [4]. The bones were found in excavations beneath medieval ruins in the ancient town of Dmanisi in the Republic of Georgia, a site that falls along the probable *H. erectus* migration route [5, 6].

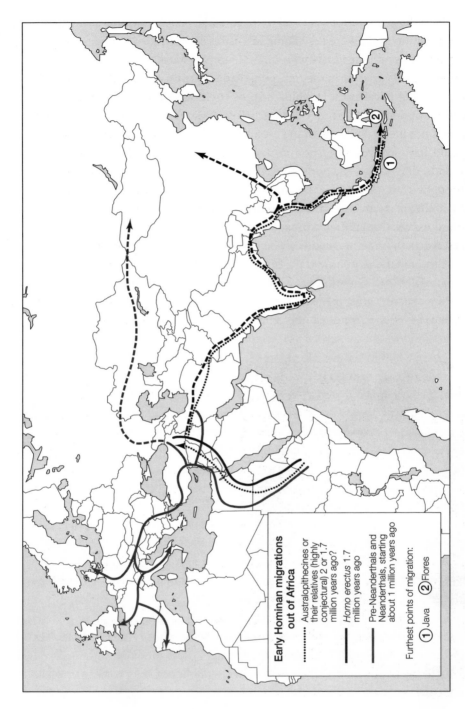

Figure 58 This map of the early migrations of our relatives out of Africa is based on both fossil and molecular data.

Were these mysterious Australopithecine-like people part of a separate earlier migration [5]? Were they companions or slaves of *H. erectus*, or were they simply unusual individuals who popped up periodically because there was a large store of variation in the gene pool of the *H. erectus* migrants? I favor the last explanation, because the Hobbits and the people of Dmanisi show a mix of Australopithecine and *H. erectus* characteristics. The large store of variation hypothesis is not incompatible with the evolution of the Hobbits—their short stature could have evolved in response to shortages of food on their small remote islands, drawing on the pool of genetic variation that their *H. erectus* ancestors brought from Africa. If so, then their evolution would have taken a course similar to that of the pygmy mammoths of Santa Cruz Island. Further fossil discoveries, especially of Hobbit bones that still contain intact DNA, will help to resolve these and other puzzles about the *Homo erectus* migration (Figure 59).

Although these *Homo erectus* migrants did not manage to venture quite as far from Africa as the later wave of modern humans, they were able to traverse huge distances across great physical barriers. Stone tools and *H. erectus* skeletons that date to between 1.6 and 0.7 million years old have been found on the island of

Figure 59 These cheerful kids and their grandmother, busy drying palm nuts in the sun, live in the Ngada tribal village of Luba on Flores. The people who now live in Flores and the other islands of the Lesser Sunda chain of Indonesia preserve some genes from modern humans who migrated out of Africa fifty thousand years ago, mixed with those of more recent migrants from East Asia. Some of the ancestors of these children lived on the island at the same time as the tiny—and now extinct—Hobbits, *Homo floresiensis*, though there is no evidence that they hybridized with them. The Hobbits, who lived on Flores as recently as 12,000 years ago, were in turn perhaps descendants of *Homo erectus* and other migrants who left Africa close to two million years ago. Did the Hobbits, too, leave some genes behind?

Java (marked with a 1 in Figure 58). On Flores, a scattering of similar stone tools in the central part of the island have been dated to between 800,000 and a million years ago [7]. But the trail of these hominan migrations ends on Flores.

In order to get to Flores, the *Homo erectus* bands would have had to cross at least three narrow ocean channels that split the Lesser Sunda island chain. The channels are scoured by fierce north-to-south currents. Were all of them crossed by half-drowned people who were swept to sea after storms, and who survived by clinging to floating logs? This seems unlikely. Perhaps, in spite of their primitive technology and small brains, the *H. erectus* migrants were able to build coracles or crude canoes.

Such simple boats may not have been capable of crossing the much wider deep-water channels that separated the Lesser Sunda chain of islands from Australia and New Guinea. Perhaps this is why *Homo erectus* never reached those remote lands. But these hominans were still hugely successful in their new environments. They survived on Java for the better part of a million years—hundreds of thousands of years longer than their relatives who had stayed behind in Africa [8]. And their possible relatives, the Hobbits, survived on Flores until 12,000 years ago [9].

Later Migrations Out of Africa

Ten million years ago, the Middle East was wet enough to allow hippopotamuses to spread to India from Africa [10]. When the first bands of *Homo erectus* ventured across this region eight million years later, the ice ages had not yet reached their full extent. The planet still retained some of its earlier warmth and abundant rainfall. Although the land was growing dryer, the *H. erectus* migrants would have found that at least parts of their route resembled their home savannahs of East Africa.

After the *Homo erectus* migration there seems to have been a long pause. As the deserts grew dryer it became more difficult for migrants to leave Africa.

We saw in Chapter Five how the great collision of the Indian and Asian tectonic plates may have triggered the repeated ice ages of the last two and a half million years. Whatever the cause, whenever the Earth cooled in the grip of a new ice age, the Middle East dried up. Its shimmering deserts posed a formidable barrier.

In spite of these difficulties, starting about a million years ago a series of migrant bands were again able to make their way out of Africa. These peoples, like *Homo erectus* before them, funneled through the Middle East (Figure 58). They scattered both east and west, colonizing the empty lands of Europe and Central Asia.

The first of these later migrants were a heterogeneous group collectively known as the pre-Neanderthals. As they spread into Europe they left fossil traces behind at sites as widely scattered as Greece, Spain, Germany, and Britain [11]. Another line on the map traces a route, leading into Central Asia, which was followed by a separate group of migrants. Nothing is known about these mysterious people, a few traces of whom have been found in a cave called Denisova in Siberia's Altai Mountains. Nothing, that is, except—as we will see shortly—their genes.

Another wave of migrants, genetically related to the Denisovans, also started from East Africa. These would become the Neanderthals, robust and big-brained peoples who colonized much of Europe and western Asia. The exact timing of the Denisovan and Neanderthal migrations out of Africa is unclear, but it is known that their ancestors diverged from our own lineage about 800,000 years ago and they diverged from each other about 650,000 years ago. These dates are probably earlier than their departures from Africa, so that they are likely to have left Africa separately.

We do not yet know whether the Neanderthals simply displaced the pre-Neanderthals that they encountered in Europe, or whether they interbred with them. The Neanderthals themselves may have survived down to as recently as 24,000 years ago [12], though the actual date of their final disappearance is uncertain and may be older [13].

The details of these population movements are likely to undergo dramatic revision in the future, as fossil and genetic evidence accumulates. But it seems safe to conclude that the migrants altered forever the ecosystems that they encountered. As early as 400,000 years ago, pre-Neanderthals in northern Europe were using sophisticated wooden spears to slaughter large numbers of wild horses [14].

The changes that these migrants made to their environments were also shaping their own history, genetics, and societies through a gene-brain-body-culture feedback loop that we will explore in Chapter Eleven. Clearer evidence for such changes comes from the most recent of the great migrations, the first to involve modern humans (Figure 60). When these migrants left Africa about 100,000 years ago, they set in motion an astonishing series of environmental, evolutionary, and cultural changes.

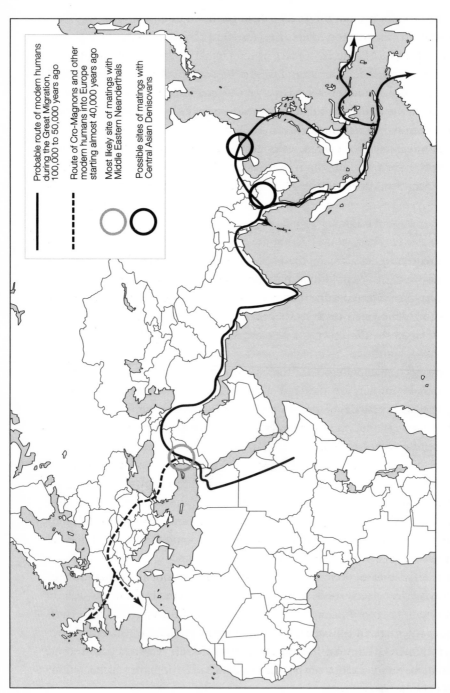

Figure 60 Small bands of modern humans embarked on their own great migration route out of Africa, starting about 100,000 years ago. They ended up in Australia and New Guinea 50,000 or more years later. The black line traces probable routes for the migration. A dark circle shows the most likely region where Neanderthal genes introgressed into our gene pool. Some of the people whose ancestors made it all the way to the end of this most recent migration also carry genes from relatives of the Neanderthals, the Denisovans. The lighter circles indicate two of the possible places where this second introgression may have happened.

As the modern human migrants spread from Africa to the farthest shores of Australasia, they encountered the Neanderthals, who were already well-established in the Middle East. Later in the course of the migration they met Denisovans, who had earlier spread to various parts of Asia [15]. And some of these Neanderthals and Denisovans mated with modern humans.

We saw in the last chapter that introgressions may have influenced the early split between the human and chimpanzee lineages, but the evidence is slender. There is much stronger evidence that introgressions from Neanderthals and Denisovans have influenced our recent history. How did these most recent introgressions happen, and what do we know about their genetic effects?

A Tryst in the Levant

As I climbed up to Tabun Cave on the slopes of Israel's Mount Carmel on a hot bright day in early 2009, I found that the Ministry of Tourism had been there before me. Cute cutouts of ape men with missing faces dotted the site, allowing visitors to pose for pictures as faux Neanderthals. Such kitch aside, other signs that marked the site did a good job of summarizing the cave's long history.

Tabun, first excavated by British archeologist Dorothy Garrod in 1929, is one of the places where modern humans and Neanderthals may have encountered each other (Figure 61) [16]. The region around Tabun may even be where these two divergent groups might—to put it delicately—have introgressed.

Tabun Cave is named after clay and stone ovens that were built there by nineteenth century goat-herders. But the cave has a far longer history. Neanderthal skulls have been found in the cave's Layer C, which is approximately 100,000 years old [17]. Stone spearpoints and other stone artifacts from the deeper Layer G show that people lived in the cave as early as 300,000 years ago. Were the first cave occupants also Neanderthals? If they were, then they must have lived in the hills and valleys that surround the cave for at least 200,000 years. Their population must have been thriving and stable.

For at least some of this time the Neanderthals were not alone. Skeletons of people who were clearly modern humans have been found in the nearby cave of Skhul, and in the more distant Qafzeh rock shelter of the upper Galilee. These skeletons resemble the oldest modern human fossils that have been discovered. These fossils, from Ethiopia, have been dated to 150,000 years ago [18]. It seems likely that the Skhul and Qafzeh people originated in Ethiopia or in other nearby parts of northeastern Africa.

Figure 61 Tabun Cave, on the slopes of northern Israel's Mount Carmel. The indicators that have been placed in the cave mark the locations of different layers of deposits, ranging from the relatively recent B down to the much older G. Each layer has been painstakingly examined by generations of archaeologists.

The modern human remains of Skhul and Qafzeh have been dated to between 140,000 and 80,000 years ago [17]. It is likely that these two lineages lived almost within shouting distance of each other, perhaps for a period of tens of thousands of years. Then, for unknown reasons, the Neanderthals disappeared as they later did in Europe. But here in the hills of northern Israel they left a genetic legacy behind.

Much else was happening to the populations of modern humans in the region. Perhaps fifty or sixty thousand years ago, a tiny group of them managed to battle their way across the deserts that blocked their route to western and southern Asia. Evidence that only a few of them made the journey can be found in the *mitochondrial chromosomes* of their descendants.

A Mitochondrial Interlude

Mitochondrial chromosomes are little snippets of DNA, about 16,000 bases long, that we all inherit from our mothers and that can tell us much about our ancestry. Actually, these little pieces of genetic material did not start out as our

DNA at all. They can be traced to distant ancestors of *mitochondria*, incredibly useful little bean-shaped structures that reside in large numbers inside most of our cells.

Mitochondria are the remote descendants of free-living bacteria that invaded our ancestors about two billion years ago. This invasion happened at a time when our ancestors, too, were simple single-celled organisms. The invading bacteria may at first have caused disease in the cells that they infected, but they soon became our mutualistic symbionts. In the process they became essential to our ancestors' survival.

The invading bacteria were able to *respire aerobically,* which means that they could extract large amounts of extra energy from their food by using oxygen in a kind of controlled burn. As they invaded, they bestowed this ability—or rather re-bestowed it—on our ancestors.

It turns out that our single-celled ancestors had also once been able to carry out aerobic respiration, but they had lost this capability. This was not careless-ness on their part. Instead, they had entered into an evolutionary Faustian bar-gain. The membranes that surrounded their cells had once been an essential component of their aerobic respiration mechanism, but these membranes had become modified so that our ancestors were able to engulf other living cells. In the process, their respiratory ability was lost. Our progenitors had traded a great metabolic advantage for the ability to become the planet's first predators.

The arrival of the later bacterial invaders canceled this Faustian bargain. When they invaded our ancestors' cells they restored their lost ability to use oxygen but did not affect their ability to hunt prey. Our ancestors continued to engulf and digest other cells. In return, the ancestors of the mitochondria got copious quan-tities of food and a fine, safe place to live [19].

After all this time, the mitochondrial chromosomes have shrunk to a tiny snippet of their former size. The information that they carry is dwarfed by the six billion bases of DNA on our other chromosomes. Nonetheless, this informa-tion is incredibly important to our survival because it helps us to burn our food efficiently. The benefits that are conferred by the mitochondria and the genes that they carry have been centrally important to the evolution of large multicel-lular animals and plants such as humans, whales, and redwood trees that depend utterly on the ability to extract huge amounts of energy from food.

Our mitochondria have accompanied us everywhere on our travels. But we only inherit our mitochondria from our female ancestors. Our fathers' mito-chondria are lost when their sperm dissolve at the time of fertilization. Mothers pass their mitochondria, along with those little chromosomes, to their offspring. The little chromosomes are thus inherited in a matrilineal fashion.

This pattern of inheritance means that the mitochondrial chromosomes of our parents do not have a chance to recombine with each other. The genes on our other chromosomes are shuffled each generation, just before parents pass their genes to their progeny. But the only way that the matrilineally inherited mitochondrial chromosomes can change over time is through the gradual accumulation of mutations.

Because of their simple pattern of inheritance, it is the mitochondrial chromosomes that tell us most clearly what happened when the first bands of

Figure 62 An Orang Asli village headman, his blowpipe over his shoulder, shows me the tiny poison-laden dart that he was able to bury in a target from fifty feet away. His tribe, which has lived in the peninsular Malaysian rainforest for tens of thousands of years, carries genes from the earliest modern human migrants out of Africa. During most of this long period the ancestors of the Orang Asli exploited their forest's rich resources in a sustainable way. Now, with the spread of agriculture, their remaining groups are running out of animal resources. Some of these people of the forest have joined the army, and others have become guides in peninsular Malaysia's largest national park, Taman Negara. [21]

modern human migrants managed to cross the deserts of the Middle East. Because mitochondrial chromosomes have independently acquired different mutations in different human lineages, we can use these mutations as markers to track our divergent genetic histories as we have fanned out across the planet.

The people who left the Middle East behind took with them only three of the many different types of mitochondrial chromosomes that were carried by their African ancestors. This fact tells us that the migrant band probably numbered only a few hundred individuals at the most. And it is likely that most of them were related to each other [20].

Once the descendants of this tiny band had managed to cross the deserts that lay beyond the Mediterranean coast and reached the lusher realms of western India, they had a much easier time of it. The paths along the Indian coast to Southeast Asia, and eventually to the islands of Indonesia, New Guinea, and Australia, were far less challenging than the deserts. Descendants of some of these original migrants are still found today in southern India, scattered through Southeast Asia, and in Australia and New Guinea. They all carry variants of those three original mitochondrial DNA types.

It is likely that, as the migrants penetrated to each new area, they promptly devastated it by overexploiting the easiest resources. A few of them would then have simply moved on to new lands, like the Mad Hatter at the Wonderland tea party who periodically shifted seats and took possession of a fresh table-setting. Those who were left behind to deal with the mess soon had to learn a more sustainable lifestyle (Figure 62).

The saga of these peoples can be followed at many levels—ecological, sociological, and genetic. In the next chapter we will trace their amazing genetic story.

....

CHAPTER EIGHT

Blending and Balance in Our Gene Pool

S cattered fossil finds, and DNA information from the people who now live in the paths of the great migrations out of Africa, have only given us a partial picture of how modern humans first spread from their ancestral continent. Now, DNA from fossils is filling in the story. We are starting to understand the genetic differences that separate us from our closest relatives. We can now probe the very nature of what makes us uniquely human, and understand how we have been able to adapt so readily to such a diversity of ecosystems around the planet.

Ancient DNA

The first DNA to be sequenced from a long-dead organism was isolated from a 2,400-year-old Egyptian mummy by Swedish evolutionary biologist Svante Pääbo in 1985 [1]. A quarter of a century later, thanks to advanced technology, it is now possible to resurrect and read short bits of DNA that are as much as 400,000 years old. Genetic information has been recovered from many ancient organisms, ranging from marsupial tigers and woolly mammoths to our extinct Neanderthal and Denisovan relatives.

Sometimes, however, even the most advanced DNA sequencing techniques can fail utterly. We would love to have DNA sequences from Asian *Homo erectus* and from their possible relatives the Hobbits. If we could read these sequences, we would learn much about the mysterious first migrations of hominans out of Africa.

For example, who were the Hobbits? Were they simply miniaturized versions of *H. erectus* who had become dwarfed because of a shortage of food like the pygmy mammoths of the California Channel Islands?[1] Or were they on a different hominan branch altogether? How genetically different were the Hobbits and

H. erectus from the Neanderthals and from ourselves? And are there any signs that genes might have introgressed from the Hobbits or from *H. erectus* into our own gene pool?

Alas, the *H. erectus* bones and the Hobbit bones have gone through changes that make it impossible to extract their DNA. Some of the finest Asian fossils of *H. erectus*, which were discovered near Beijing in China in the 1920s, disappeared at the outset of World War Two. We are left with nothing but casts of the original bones. The *H. erectus* fossils from Java are too infested with tropical bacteria and molds to harbor any of the DNA that belonged to their original owners. And even though the bones of the Hobbits are only a few tens of thousands of years old, they have been soaked with water for so long that their DNA has also vanished.

But, amazingly, sequences have been obtained from Neanderthal and Denisovan DNA. These fossils have been preserved under colder and drier conditions. Their sequences show that there was introgressive gene flow from these earlier migrants into some of us. The discovery of these introgressions opens up the possibility that our lineage might also have been the beneficiary of similar gene flows that took place in our more distant past.

The discovery of these introgressions was unexpected, because a substantial amount of earlier genetic data from Neanderthals had seemed to rule them out.

The first Neanderthal DNA was sequenced in 1997, by Matthias Krings, a graduate student in the Leipzig lab of Svante Pääbo [3]. He extracted the DNA from an arm bone of the most famous Neanderthal fossil, the 40,000-year-old "type specimen" that had been discovered in 1856 in Germany's Neander Valley [4].

The technology that Krings used now seems quaint. The amount of DNA that he obtained was laughable—fewer than 400 bases. And it came from the Neanderthal's mitochondrial chromosomes, which are found in dozens or hundreds of copies in each cell and is therefore the easiest old DNA to find. Nonetheless, even though this Neanderthal sequence was short, it showed some clear differences from the corresponding sequences of modern humans.

Because of these differences, Krings concluded that Neanderthal mitochondrial sequences were unlikely to have survived down to the present time in our species. His conclusion has continued to stand in the fifteen years since. Thousands of modern human mitochondrial chromosomes from every part of the world have been compared with Neanderthal sequences, and there are still no matches [5, 6].

Just because such chromosomes have not been found, however, does not mean that human–Neanderthal matings never took place. If a few mitochondrial chromosomes from Neanderthals had entered our gene pool tens of thousands of years ago, they could easily have been lost. Male hybrids would not have passed the chromosome to their children, and the mitochondrial chromosomes of female hybrids might have been lost by chance in subsequent generations.

Other Neanderthal genes, however, would not have been so easy to lose after hybridization. These genes are coded in the 23 chromosomes that are located in the nucleus of the cell. They carry the *nuclear DNA*, which contains more than 200,000 times as much genetic information as the tiny mitochondrial chromosomes.

If such Neanderthal nuclear genes had introgressed into our ancestors' genomes, they would have originated from many different Neanderthal chromosomes. And they could have been passed to subsequent generations from male as well as female hybrids, because nuclear genes are passed on by both sexes. Even if the mitochondrial chromosomes from an introgression were all lost, nuclear genes could have persisted.

Svante Pääbo, Richard Green, and their group of paleomolecular detectives at Leipzig's Max Planck Institute, along with David Reich, Nick Patterson, and their colleagues at M.I.T.'s Broad Institute, have taken advantage of improving DNA sequencing technology. They turned again to old Neanderthal bones in a hunt for these more elusive nuclear genes. They began with DNA from bones of three female Neanderthals who had died 40,000 years ago and whose remains had ended up in a cave in Croatia.

Such ancient DNA poses enormous challenges. Most of the DNA in ancient bones comes from bacteria that have invaded after the bones' owners died. Luckily, these sequences are so different from human or Neanderthal DNA that they can be picked out and discarded. More problematically, many of the Neanderthal DNA fragments have become chemically modified over time. Luckily, many of these changes can also be detected and edited out.

At the end of an extensive editing process, the bone extracts yielded sequences from millions of tiny Neanderthal fragments [7]. Most of this DNA was nuclear, not mitochondrial, and most of it really was from the ancient Neanderthal females. There was virtually no trace of Y chromosome DNA or of modern human mitochondrial DNA that might have come from museum workers who had handled the bones.

They also looked intensively at differences between replicate reads of the same sequence of DNA that had been obtained from the same Neanderthal bone. These differences might have indicated modern DNA contamination. But when these variant reads were compared with modern human DNA, they almost never matched the modern sequences. This suggested that they were actually ancient Neanderthal genetic polymorphisms [8].

Green's group was able to match up about a third of the entire genomes of Neanderthals and modern humans. As expected, they found fascinating information about the ways in which individual Neanderthal and modern human genes have diverged during the four hundred thousand years since their common ancestor. But they also, unexpectedly, found that between one and four percent of the nuclear genes of present-day Europeans and Asians can be traced to Neanderthals. There was no sign of such introgression in the genes of modern sub-Saharan Africans, whose ancestors would not have encountered any Neanderthals.

How could they be sure of the introgression signal? Unlike a mitochondrial chromosome, which is inherited as a unit, any introgressed DNA from Neanderthal nuclear chromosomes would have been broken up and interspersed with modern human DNA through a thousand generations or more of genetic recombination. To find signs of such introgressed Neanderthal fragments, it was necessary to plod through the human and Neanderthal sequences and compare them base by base.

Going Gaga over ABBA-BABA

Green's group used two clever and path-breaking methods to look for introgression. The methods gave similar but complementary results.

Any one of four possible molecules, referred to in chemical shorthand as bases, can occupy a given site along one strand of the DNA double helix (Figure 63). These molecules are adenine, thymine, guanine, and cytosine, and they are commonly abbreviated as A, T, G, and C. The way in which the bases are arranged along the strand encodes the DNA's genetic information. Each amino acid is encoded by a sequence of three bases. For example, if a bit of DNA at the start of a gene has the sequence ATGGTG, we can predict that the protein the gene codes for will begin with the amino acids methionine and valine.

The sequence of DNA bases in our genomes ultimately determines the structures of our billions of component cells, and directs how each of us has grown from a single cell into an adult human. Ninety-nine percent of these long sequences of bases do not code for proteins. Nonetheless, the information carried by this non-coding DNA can be important in determining how genes are regulated and how chromosomes fold and replicate.

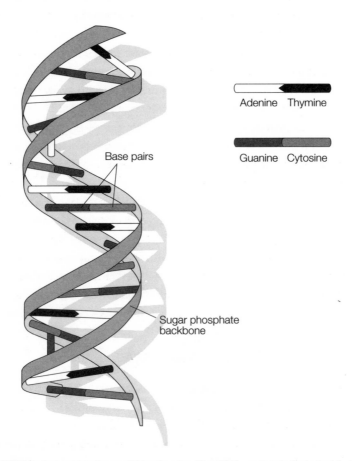

Figure 63 DNA is the centerpiece of this chapter. The DNA molecule is a double helix of sugars and phosphate molecules, which form a scaffold for a ladder of paired molecules that are given the shorthand term of bases. Cradled within the framework of this double helix, the base adenine always pairs with the base thymine, and guanine pairs with cytosine. The sequence of bases along the rungs of the ladder carries the DNA molecule's genetic information. When DNA replicates, enzymes pull it apart into two single strands. Cellular machinery synthesizes new complementary strands, and the result is two daughter molecules that are identical to the original. The two strands of DNA can also be pulled apart temporarily, allowing one of the strands to serve as a template for the synthesis of complementary single-stranded messenger RNA molecules. The messenger RNA carries the DNA's genetic information from the nucleus to other parts of the living cell.

The first test for introgression that Green and his colleagues used was endearingly named ABBA-BABA, which sounds like a cross between a Swedish rock group and a character from the Arabian Nights. The ABBA-BABA test depends on the fact that small mutational changes have accumulated in the DNA sequences of ourselves and our relatives as we have diverged over the generations.

Any of the bases in a sequence of DNA can mutate into any of the other three. For example, if A mutates to C, then in the daughter DNA molecules the AT pair is changed into a CG pair. Such tiny mutations are similar to the errors that crept into medieval manuscripts as they were copied again and again.

The ABBA-BABA test constructs little family trees for each site along the DNA molecules, and looks for patterns that are generated by certain types of mutational change. These little family trees tell many stories. Figures 64 through 66 show some of them.

The top of each family tree in the figure shows the base that was probably present at this site in the common ancestor of humans, Neanderthals, and chimpanzees. Although we can infer what the base probably was, we cannot be sure because this ancestor lived between six and seven million years ago.

At the bottom of each tree are the bases that are found at this same DNA position in corresponding sequences of present-day human, chimpanzees, and Neanderthals. In each tree, the labels H1 and H2 indicate sequences that have been obtained from two groups of modern humans who live in different parts of the world.

The X shows where and when a mutational change probably happened. Tree 1 of Figure 64 shows an early change from a T to a C, which altered the base on the other strand of the double helix from an A to a G. In this tree the change is shown as taking place in the chimpanzee lineage, but it is also possible that a C to T change might have happened early in the human–Neanderthal lineage. Tree 2 of the figure is less ambiguous: there clearly was a change from a G to a T in the human lineage after it diverged from the Neanderthals.

In tree 1, chimpanzees have a C and the human and Neanderthal sequences all have a T. This type of tree can be described by a diagnostic pattern that is shown on the bottom line. In this case the pattern is AAAB, and it would be the same in similar trees regardless of which bases happen to be involved. In tree 2, the pattern is AABB.

Note that neither of these tree patterns reminds us of a rock group or a character from the tales of Scheherazade. Trees like the ones in Figure 64 turn up often in the data. But if they don't form an ABBA or a BABA pattern, the test ignores them.

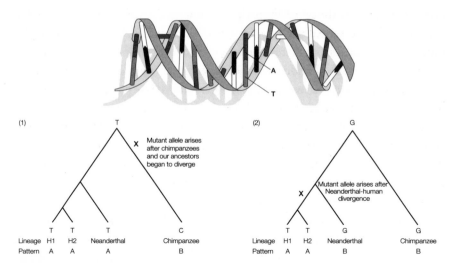

Figure 64 Two common types of the many different family trees that Green and his colleagues found. The tree on the left follows the fate of a T (for thymine) at a particular site in the DNA sequence of the common ancestor of humans, Neanderthals, and chimpanzees. In this tree it is supposed that an ancestral T mutated to a cytosine (C) in the chimpanzee lineage, though it is also possible that an ancestral C mutated to a T early in the human–Neanderthal lineage. The tree on the right follows the fate of a G (for guanine) in another part of the common ancestor's DNA. This base mutated to a T in the human lineage, subsequent to the human–Neanderthal split.

A small minority of the trees do form an ABBA or BABA pattern. There are two ways in which this could happen. The first of them is shown in Figure 65.

The trees of Figure 65 give us an important clue to how the split between the ancestors of humans and Neanderthals took place. As these two groups' gene pools gradually diverged they carried with them pre-existing genetic differences between individuals. Sometimes mutant alleles persisted in at least one of these diverging gene pools for a while, without taking over completely. During this time the new mutant allele and the old allele of the gene coexisted. This site of the DNA was, to use genetic terminology, polymorphic.

Occasionally, the old allele of the gene survived for so long that it was passed down the Neanderthal lineage and to one of the human lineages, while the new mutant allele was passed down to the other human lineage. The figure shows how such an event would give rise to an ABBA or a BABA tree, depending on which human lineage received the new allele.

The existence of such trees may seem counterintuitive—how can a new mutant allele that arose at that ancient time, more than half a million years ago, be passed down to only one of the two different human lineages? Surely, the two human lineages must have diverged much more recently, and should therefore have carried only one of the alleles. But there is strong evidence that alleles can sometimes survive in populations for hundreds of thousands or even millions of years without either being lost or taking over the entire population. It is just such persistent polymorphisms that can generate the ABBA or BABA trees of Figure 65.

The trees of Figure 65 can happen in the absence of any introgression between Neanderthals and humans. And, importantly, we would expect equal numbers of both types of tree. The detection of introgression depends on the fact that there is another way to generate ABBA or BABA trees, one that produces unequal numbers of the two types of tree. These cases are shown in Figure 66.

The heavy arrows in Figure 66 show how a Neanderthal base can be passed by introgression to only one of the two human groups. If Neanderthal bases were preferentially passed to human group H2, the result will be an excess of ABBA trees. If the Neanderthal bases were instead passed to human group H1, the result will be an excess of BABA trees.

Green and his coworkers found that if the first human group H1 came from sub-Saharan Africa and the second group H2 came from Europe or Asia, there was an excess of ABBA trees among the thousands of little trees that they examined. In the reverse case there was an excess of BABA trees. The pattern of these excesses showed that there had been introgression of genes from Neanderthals into the European and Asian gene pools, but not into the sub-Saharan gene

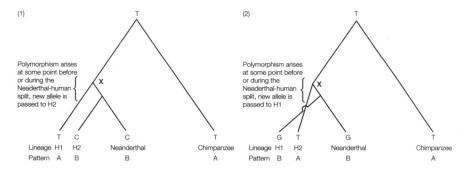

Figure 65   These trees, which yield an ABBA or BABA pattern, are relatively rare because they require that the mutation happened during the time of the split between Neanderthals and humans. The mutation remained polymorphic during the time that was encompassed by the brackets in the figure. It survived during this period in such a way that it was passed down the Neanderthal lineage and only one of the human lineages. Each tree is equally likely to happen, so that these ABBA and BABA trees should be equally common.

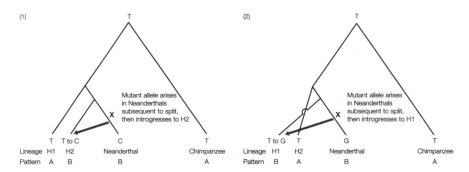

Figure 66   If a mutation arises in the Neanderthal lineage subsequent to the human–Neanderthal split, and the mutation is then passed to one of the two human groups by introgression, this will produce additional ABBA or BABA trees over and above those that we saw being generated in Figure 65. Depending on which human group receives the introgressed genes, there will be an excess of ABBA or BABA trees.

pools. They could calculate that between one and four percent of the DNA of Europeans and Asians had introgressed from Neanderthals.

Because Europeans and Asians showed the same amount of introgression, there was a strong likelihood that the human–Neanderthal matings had happened before European and Asian humans parted company. The most likely time and location was when the first modern humans arrived in the Middle East from Africa and encountered Neanderthals, perhaps in a region near Tabun Cave. If modern humans had mated again with other populations of Neanderthals as they moved into Europe or western Asia, then the people living in these regions today should carry more Neanderthal genes than modern humans elsewhere. Such a pattern is not seen.

My explanation of the ABBA-BABA test has glossed over some complexities in the analysis, such as the rare cases in which two or more mutations happened in the same tiny family tree. But these complexities can be accounted for without disturbing the conclusions from the test. This analysis really does show that at least some of the Middle East encounters between modern humans and Neanderthals were anything but warlike, and instead enriched our gene pool.

How Much of the Neanderthal Genome Survives in Us?

Luckily, the conclusion that Neanderthal genes have introgressed into our gene pool does not rest entirely on the ABBA-BABA test. We would hesitate to reach such an important conclusion if we had only a single type of evidence. Green's group has also analyzed the human and Neanderthal genomic data in a different way. Their second analysis reached the same conclusion as the ABBA-BABA test about the reality and the magnitude of the introgression. And, although they did not discuss it, their second test opens the door to a fascinating and important question about the role that Neanderthal genes play in those of us who carry them.

Every present-day European and Asian carries between one and four percent of Neanderthal DNA sequences. But do we all share the same limited set of pieces of surviving Neanderthal DNA? Or do different present-day humans carry different parts of those ancient genomes? Perhaps, if we were to scan the genomes of many different people living today, we would be able to piece together a substantial part of the Neanderthal gene pool scattered about in the form of different bits of Neanderthal DNA that have been preserved in different modern human lineages.

The ABBA-BABA test cannot answer this question. The excess ABBA or BABA trees of Figure 66 serve as signals of introgression, but they are scattered thinly along the chromosomes. And because they are interspersed among the more numerous ABBA and BABA trees of the type shown in Figure 65 that were not produced by introgression, there is no way to distinguish which are which.

In the second test, Green and his colleagues compared the Neanderthal DNA sequences to DNA from two modern humans. The first was Craig Venter, a biologist of European ancestry, and the second was one of the people whose DNA was completely sequenced by the National Institutes of Health (NIH) sequencing project.[2] About seventy percent of the NIH genome comes from a single individual with half European and half African ancestry, and it was this individual's genome that they used in their comparison.

They began by breaking the modern human genome data down into small pieces of DNA sequence that were tens of thousands of base pairs long. They then matched these pieces with the equivalent pieces from Neanderthals.

Like different drafts of a manuscript, these matching human and Neanderthal pieces resembled each other, but not perfectly. Some of the matched pieces differed at many sites along their sequences, showing that these regions of human and Neanderthal chromosomes had evolved rapidly away from each other. Other regions resembled each other much more closely, indicating that little evolutionary divergence had taken place.

Such comparisons were not enough to show an exchange of genes between humans and Neanderthals, however. Because different parts of genomes evolve at different rates, a close resemblance between a human and a Neanderthal sequence could simply mean that this part of the genome happens to evolve slowly.

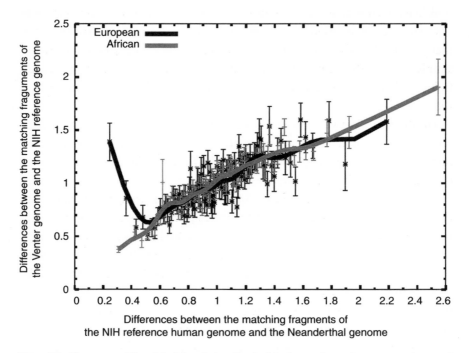

Figure 67 The upward "hook" of the darker line in this figure shows how some pieces of Craig Venter's genome have a weaker resemblance to Neanderthal DNA than the corresponding sequences from the European half of the NIH genome. The hook is not seen when the African-American half of the NIH genome (lighter "African" line) is compared to the Venter genome. (From part of Figure 5 of [7].)

To look for signs of introgression, they tracked down pieces of the European part of the NIH genome that resembled the corresponding Neanderthal pieces closely. They then examined the equivalent pieces in the Venter genome, to see whether they also resembled the Neanderthal pieces (Figure 67).

If all three pieces resembled each other, then this was simply a slowly-evolving part of the genome. But if the NIH piece resembled the Neanderthal piece and the Venter piece did not, this indicated introgression. The NIH genome had received a piece of Neanderthal DNA that had been preserved during the intervening tens of thousands of years, while the Venter genome had not received the same piece. You can see this effect as an upward hook at the left end of the darker ("European") line in Figure 67. The hook indicates that there was a mismatch between some of the Venter and NIH pieces.

When the same analysis was done using the African-American regions of the NIH genome, it was found that whenever they resembled pieces of the Neanderthal genome they also closely resembled pieces of the Venter genome. These are simply parts of the human genome that have in general evolved in a leisurely fashion. There is no Neanderthal signal buried in the African-American part of the NIH genome.

The difference between the European and the African-American parts of the NIH genome is to my mind the most powerful evidence for matings between humans and Neanderthals. The African-American results constitute a strong control, a control that is built right into the data.

The most obvious artifact that could ruin these analyses is contamination of the Neanderthal genomes by modern European DNA. As we have seen, several lines of evidence show that this did not happen. But if it had, and if it had been responsible for the "hook" signal in Figure 67, then some of the "Neanderthal" sequences would actually have been sequences from modern humans and the clear distinction between the European and African-American halves of the data would have been masked by the contamination. The shape of the signal in Figure 67 can only be explained by Neanderthal introgression.

The results of the second analysis show that the NIH genome has pieces of Neanderthal DNA that are not found in Craig Venter's genome. It is a good guess that the reverse is also true. These little pieces of Neanderthal genome have managed to survive in different lineages of modern humans for fifty thousand years or more, a period of time that spans twenty-five hundred human generations. During all this time they have neither been lost nor spread to complete fixation through the entire Eurasian gene pool.

Surveys similar to the tests of Figure 67 will soon track down how much of the Neanderthal genome survives in different Europeans and Asians. Is it only a small fraction, or do genes from all parts of the Neanderthal genome still survive somewhere in the Eurasian gene pool?

And did these bits of Neanderthal DNA survive simply by chance? Or have they reached genetic equilibria in our species because of a balance of selective pressures? If such balanced equilibria can be demonstrated, they would be reminiscent of the complex green equilibria that we have encountered repeatedly in ecosystems and gene pools around the world. Later we will see growing evidence that our gene pool has been subject to balancing evolutionary pressures with similarities to those that have shaped both the ecosystems and the gene pools of other animals and plants around the world.

A Second Tryst During the Great Migration

Given these new results from the DNA of ancient bones, the Neanderthal remains that Dorothy Garrod found in those caves in northern Israel suddenly take on a greater depth of meaning. It is likely that humans and Neanderthals really did live together on the slopes of Mount Carmel, and really did interact with each other.

I would like to imagine that the introduction of Neanderthal genes into the modern human gene pool happened as a consequence of some fire-lit scene of interracial drunken good-fellowship. Alas, the more likely scenario is that some modern humans captured Neanderthal women and raped them, and that the resulting children became part of the tribe.

The Middle Eastern Neanderthals eventually disappeared, but we are not sure when this happened because their fossil record is fragmentary. Ominously, several thousand years after modern humans arrived in Europe, the Neanderthals who had lived on that continent for hundreds of thousands of years also vanished, apparently without further genetic exchange. Why were there no further genetic exchanges as modern humans encountered these western Neanderthal populations? Did the modern humans arriving in Europe bring diseases that decimated the Neanderthals? Were the encounters of European Neanderthals with modern humans so fierce that no exchanges were possible? We cannot at the moment decide among such scenarios, but the stunning fact remains that introgression from Neanderthals did happen, probably in the Middle East.

It now turns out that the Middle East encounter was not the only one between modern humans and the peoples who had gone before them. Early in 2010, Pääbo and his group reported that they had extracted DNA from a Denisovan. The DNA came from a single finger bone of a child who had died thirty to fifty thousand years ago in the Denisova cave of central Siberia [10].

When it was found, the finger bone had been assumed to belong to a Neanderthal. Excavations in the cave had already shown that people were living in this remote part of central Asia fifty thousand years ago, and that they had used tools that resembled those used by Neanderthals. This in itself was a remarkable finding, because no one had suspected that Neanderthal-like peoples had penetrated so far east. But the child to whom the finger bone had belonged was not a Neanderthal. Her mitochondrial DNA was clearly different from the equivalent mitochondrial chromosomes of either Neanderthals or modern humans.

Pääbo's group showed that the little girl's mitochondrial DNA had diverged from both human and Neanderthal mitochondrial DNAs a million years ago—long before there were either modern humans or Neanderthals! With the aid of newer DNA sequencing technology, they were soon able to explain this puzzle. They subsequently extracted enough nuclear DNA from the little girl's finger bone to yield a substantial amount of sequence from her nuclear genome. She turns out to have been related to the Neanderthals after all—though on a separate branch that parted company from the ancestry of the European Neanderthals about 650,000 years ago [11].

Her puzzling mitochondrial DNA might have been a fluke. It might by chance have been preserved all the way down from her million-year-old ancestors. An alternative possibility is that there was some introgression between the Denisovans and another group of early humans who carried older mitochondrial chromosomes. And, to make things more complicated, the cave also turns out to have been occupied by Neanderthals—a toe bone from the cave deposits, dated to approximately the same time as the Denisovan bone, contains Neanderthal DNA [12]!

In the space of a year, the little girl who died in that remote central Asian cave has taken her rightful place on a newly-discovered branch of the human family tree, the Denisovans.

This was not the end of the story, however. When Pääbo's team compared the little girl's DNA with that of modern humans, they discovered that about four percent of her sequences are found in people from Papua New Guinea [11]. More recently, Denisovan DNA has been found in an Australian aboriginal genome and in an aboriginal group in the Philippines, showing that some Denisovan genes have also spilled over into these peoples of the Pacific (Figure 55) [13, 14].

Other groups who are descended from the first modern human migration, the Negrito people of the Andaman Islands and the peoples of peninsular Southeast Asia and western Indonesia, do not carry DNA from the Denisovans. This suggests that the encounter (or encounters) of modern human migrants with the Denisovans happened, not anywhere near the cave of Denisova, but somewhere far to the east near the end of the modern humans' great migration.

These new molecular discoveries have upended the worlds of paleoanthropology and human evolution. It is now clear that our history does not consist of tidy separate branches radiating from the trunk of a well-behaved tree of ancestry. Instead, it looks more like a road map that is dotted with many different freeway interchanges. As different human groups encountered each other at these interchanges, they sometimes exchanged genes. Then they sped away again in different directions. Many of these directions led towards eventual extinction, but our own lineage survived and carries signs of these ancient encounters.

Our gene pool has been influenced by these introgressions, and probably by earlier introgressions that took place among various groups of hominans when all of our ancestors lived in Africa. As I was completing this book, some tentative evidence emerged for such earlier introgressions. Sarah Tishkoff of the University of Pennsylvania and her colleagues have found tantalizing glimpses of unusually old bits of DNA in the genomes of three ancient lineages of hunter-gatherers in different parts of Africa—the Hadza and the Sandawe of Tanzania, and the West African pygmies [15]. These old bits of DNA are mostly shared by all three of these populations of hunter-gatherers, and have been preserved during the time that these populations have themselves diverged substantially from their common ancestor. These ancient bits of DNA are as different from modern human DNA as is the DNA of Neanderthals, though they do not resemble Neanderthal DNA.

The picture that Tishkoff and her colleagues present is one of ancient introgressions between the ancestors of some modern human lineages in Africa and a group of people, as yet undetected in the fossil record, who were as divergent from us as Neanderthals. Their results are tentative, because they cannot make direct comparisons between modern sequences and the sequences of these ancient people. But the statistical method that they use appears to be valid, because when it is applied blindly to modern European DNA it fishes out Neanderthal sequences.

With each of these new discoveries, our genetic history grows in complexity. The genomes of the people who stayed in Africa, like the genomes of those who left, have been shaped by gene flow and subsequent natural selection. As we will see in the next chapter, it is likely that these multiple introgressive gene flows have added to the adaptability of our species as we have invaded new and challenging environments.

....

CHAPTER NINE

The Intertwined Histories of Humans and Their Ecosystems

The history of our species is a history of adaptation, but it is more than that. We have forced profound changes on the parts of the world that we have invaded, and they have also changed us. In this chapter we will explore some of these stories of adaptation. All of these stories have one inescapable bottom line: we are a part of the natural world, and we must learn to join forces with it rather than simply manipulate it.

Adaptation to a Terrifying Environment

In two trips in 2009 and 2011, Liz and I explored some of the most extreme ecosystems of the Himalayas. In Bhutan we traveled from the Indian border, where the Himalayan foothills begin, to some of that remote country's high valleys. In Nepal our trips ranged from the lowland tropics to the trail that leads to the Everest base camp. Everywhere, we found indications that the peoples of these regions have adapted—and are continuing to adapt—to local conditions. Some of those conditions have posed enormous challenges.

The Inner Terai lowlands, which surround Nepal's Chitwan National Park, are steamy and subtropical. In September 2009 we visited the Baghmara Community Forest outside the park's boundaries. The local nascent ecotourism operation is thriving (Figure 68). But even the new sanitized world of the Terai poses great risks. The father of our guide Raj Kumar Mahato had been killed by a rhino some years earlier. And during our visit, a Chinese tourist was killed by a domesticated female elephant. His transgression had been to feed the tired and irritated elephant what she perceived to be an insufficient number of bananas [1].

The Terai is home to a mix of peoples. Half of them, like our guide, are indigenous Tharu and half are newcomers from other parts of India and Nepal. The

Figure 68 A one-horned rhino, *Rhinoceros unicornis*, browses in Nepal's Chitwan National Park while her baby seeks more highly processed nourishment. Endemic tropical diseases have driven the rapid adaptation of the people who live in this fiercely challenging area.

Tharu have lived in small numbers in the Terai for millennia (Figure 69), and are now playing an important role in the drive for ecotourism.

Until just a few decades ago, the Tharu suffered from murderously high levels of malaria and of kala-azar, the South Asian form of leishmaniasis. Kala-azar causes hideous leprosy-like disfigurement, and can fatally infect internal organs. These diseases of the Terai also infected many members of the expeditions that surveyed the Himalayan region for the East India Company in the nineteenth century [2].

Before the 1960s, the flat, steamy, and river-laced Terai was thick with swarms of mosquitoes that carried malarias, both the deadly *falciparum* and the less severe but still extremely debilitating *vivax*. Malaria parasites spend part of their life cycle in humans. There they dine on the hemoglobin molecules in their hosts' red blood cells. Unchecked, the parasites can cause debilitating bouts of fever, and in the severest cases they can damage their hosts' brains, livers, and kidneys. Yet the Tharu themselves were almost free of malaria, even though most of them had the parasites in their blood.

Figure 69 A Tharu villager herds water buffalo in Nepal's semitropical Terai. He is genetically adapted to live in an area that until recently was shunned by outsiders because of its high incidence of malaria and of the disfiguring disease kala-azar.

The prevalence of malaria has driven the evolutionary adaptation of the Tharu. This adaptation has centered on the hemoglobin in their red blood cells.

Hemoglobin molecules, which carry the oxygen in our blood, are made up of two alpha and two beta subunits. These subunits are specified by the corresponding alpha and beta genes. Almost all of the Tharu villagers are homozygous or heterozygous for mutant alleles of one of the two alpha-hemoglobin genes on chromosome 16. In the mutant alleles much of the DNA sequence of the gene itself has been deleted.

Because of these deleted genes the Tharu make too few alpha protein subunits, so that some of their hemoglobin molecules are abnormal. The result is a mild anemia called α-thalassemia, a condition that luckily has few harmful effects. But in the presence of malaria, α-thalassemia bestows a huge benefit [3]. The members of the Tharu population who are homozygous or heterozygous for the mutant alleles—that is, the overwhelming majority of them—are protected against the severest symptoms of both the *falciparum* and the *vivax* malarias. Although many different mechanisms have been proposed to explain this protection, the story is not yet fully understood [4, 5].

The frequency of α-thalassemia alleles in the Tharu population has reached eighty percent, by far the highest that has been recorded for this condition for any human population [6]. When DDT spraying was introduced in the 1960s, malaria was largely wiped out in the Terai. But it is starting to return, which means that the Tharu are once more benefiting from their genetic shield of α-thalassemia.

A group of Italian scientists has analyzed the mitochondrial and Y chromosomes of the Tharu, in order to trace the history of these early colonists of the region. They found that some of the Tharu ancestry stretches back to the original migration of modern humans out of Africa (and of course the Tharu carry Neanderthal genes as well). But their gene pool also includes contributions from later migrants who came from Western Europe, central India, and East Asia [7].

When the ancestors of the Tharu began to colonize the Terai thousands of years ago, their diverse pool of genetic variation stood them in good stead. The colonists carried several different α-thalassemia alleles, some of which probably originated in other South Asian populations [8]. Lucio Luzzatto, one of the authors of the original study on the peoples of the Terai, told me that the Tharu may also carry a unique α-thalassemia mutation that has arisen by genetic recombination within their population.

The Tharu were lucky to have these genetic resources. Because the α-thalassemia alleles are relatively benign, they have spread almost unimpeded in the Tharu gene pool. And they are far less harmful than many other types of mutational change that confer resistance to malaria in other human populations.

One of the most harmful of these mutations is the sickle-cell mutation. This mutation affects the gene that specifies the other (beta) subunit of the hemoglobin molecule. Sickle-cell alleles are common in Africa and parts of the Middle East. People who have inherited a mutant sickle cell allele from one parent and a fully functional allele from the other are heterozygous for *sickle-cell anemia*. These heterozygotes are resistant to the severe effects of *falciparum* malaria—but their resistance comes at a terrible cost to other members of their population. Before the advent of modern medicine, children who inherited copies of the mutant allele from both parents, so that they were homozygous for the sickle-cell allele, invariably died in childhood.

As selection by the malaria parasite drove this two-faced allele up in frequency in African and Middle Eastern populations, the growing numbers of heterozygotes for the allele were protected. But the numbers of homozygotes who died tragically in childhood increased as well. Eventually, when the benefits to the sickle-cell heterozygotes were balanced by the deaths of the young homozygotes, the frequencies of the mutant allele could rise no further.

A few sickle-cell alleles have been found in the Tharu population. These alleles have not increased in numbers, as they did in Africa, because the α-thalassemia allele that the Tharu carry is so much less harmful and so much more protective against both kinds of malaria. Because of their genetic heritage, the Tharu have managed to escape the sickle-cell curse that burdens West Africans and their descendants in the Americas to this day.

The Tharu have not pulled off a similar genetic escape from kala-azar, another disease caused by protozoan parasites. Nonetheless, the disease has protected them.

Kala-azar is carried by sand flies, which swarm on the banks of the Terai's many rivers. The disease, which infects 100,000 people each year in South and Southeast Asia, is still common among the Tharu. Now, however, it is usually spotted in its early stages and treated [9].

Although kala-azar progressively disfigures the faces and bodies of its victims, they can still survive and live for years. This means that any natural selection for alleles that confers resistance to this disease has been far weaker than selection for resistance to the often-fatal malarias.

The spraying programs of the 1960s also controlled the sand fly populations, but before that time the ravages of kala-azar marked out the Tharu as surely as if they were lepers. Their deformities, which were perceived by people outside the region as a supernatural curse, served as a vivid warning to potential invaders that settling in the Terai might not be a good idea. The Tharus' malaria resistance, together with the visible stigma of kala-azar, enabled them to farm their land undisturbed for thousands of years.

Blood and Milk

The blending of many different gene pools in the Tharu provided a rich source of allelic variants. Starting with this blend, natural selection has increased the frequencies of the most effective malaria resistance alleles. These extensive genetic resources helped the Tharu to colonize the rich but dangerous lands of the Terai.

As modern humans have spread throughout the world, other types of selective pressure have also raised the stakes on the migrants' survival. In some cases, evolutionary and cultural changes have worked together in a kind of symbiosis.

171

Descendants of the first band of modern humans to cross the Middle Eastern deserts eventually penetrated to the richer lands of India and Southeast Asia. There they hunted the herds of wild banteng and gaur, but there is no indication that they ever tamed them. Neither of these species is readily tamed, and neither is a good milk producer. Gaur have been domesticated in Bhutan and elsewhere, but they are far more temperamental than domesticated cattle and are not used for milk production [10].

The later waves of migrants who headed west and fanned out into the thick forests of central Europe encountered another species of cattle, the large and fierce aurochs *Bos primigenius* (Figure 70). These animals too were hunted for millennia. But starting approximately 7,500 years ago, through what were probably insane acts of bravery, the proto-farmers of central Europe succeeded in domesticating some of the aurochs [11]. Generation after generation of selection eventually produced the present-day strains of beef and dairy cattle, *Bos primigenius taurus*.

The first aurochs to be domesticated must have had a less savage attitude towards confinement than the average in their wild herds, which allowed

Figure 70 This painting of an aurochs from the cave of Lascaux was made 17,000 years ago, by an artist who observed these formidable animals long before their descendants were tamed.

them to be raised for meat. But some of these domestic strains also produced abundant milk, and some of these in turn could be selected to produce milk even when they were not nursing their offspring.

Initially, most of these proto-farmers could not digest this new abundant source of food. But its availability brought new selection pressures to bear on the proto-farmers themselves.

As babies we survive chiefly on milk. The cells that line our small intestine are able to secrete the enzyme lactase directly into our digestive tract. When lactase encounters molecules of the milk sugar lactose, it splits each of them into two smaller molecules, the sugars glucose and galactose. We can easily absorb both of these smaller sugars into our bloodstreams, but we have no cellular machinery that allows us to absorb the unmodified lactose.

Without lactase we could not digest our mothers' milk. But in about two thirds of humans the ability to secrete this enzyme is turned off after weaning, a trait that we share with most other mammals [12]. To make matters worse, if lactose non-secreters drink milk, the undigested lactose stays in their guts and opportunistic bacteria gleefully eat it. The result can be severe gastric problems and life-threatening diarrhea.

The remaining one-third of humans continue to secrete lactase through life, which means that they are able to digest milk even as adults. At least four mutant alleles are responsible for this ability. All of these mutant alleles are small genetic changes on chromosome 2, more than 10,000 bases "upstream" from the lactase gene itself. One of these alleles is found in much of Europe and in milk-utilizing populations in India [13]. The other three are found in East Africa and in parts of the Middle East.

There are two possible reasons for the high frequency of the lactose-tolerance alleles in these populations. One is that, by chance, the alleles were already common in places where cattle were domesticated and milk first became an important part of the adult diet. The second is that the mutant alleles were initially present in only a few copies, and began to increase in frequency through natural selection after the domestication of cattle.

Evidence from ancient DNA provides support for the second scenario. Eight humans who lived in different parts of Europe between 8,000 and 9,000 years ago carried the lactose intolerance alleles that are found in two thirds of us and that turn off the secretion of lactase [14]. So did the famous Ice Man, the 5,300-year-old frozen hunter from the Tyrol (Figure 71) [15]. The European

Figure 71 Five thousand three hundred years ago, a European hunter died of arrow wounds in the high Alps when he was about 46 years old. His body froze and was soon covered with ice. This reconstruction of his appearance, based on frozen tissue, skeletal remains, and preserved hair, was made by Dutch forensic artists Adrie and Alfons Kennis. Unlike most present-day Europeans, the Ice Man was lactose intolerant as an adult. But during his lifetime European populations were already undergoing evolutionary change towards lactose tolerance.

lactose tolerance allele must therefore have been pushed by natural selection to high levels in these populations after the domestication of cattle.

Selection for lactase persistence in African agriculturalists must have taken place over the much shorter period, starting about 3,000 years ago, when cattle were first domesticated on that continent [16].

It is possible to calculate the strength of the natural selection that led to the increase in lactose tolerance. As the mutant alleles spread through the European, East African, and South Asian populations, they dragged linked stretches of DNA with them. New DNA sequencing technology now allows accurate measurement of the sizes of these pieces of DNA that are linked to the gene. If the pieces are large there has been no time for genetic recombination to break them up, which means that selection has probably been strong and recent. If they are smaller, selection has been slower and weaker, giving recombination more time to nibble away at them. The pieces surrounding the mutant alleles are large, showing that selection for lactose tolerance was indeed strong. Each generation, people carrying the lactose tolerance alleles were between one percent and an astonishing fifteen percent more likely to survive than those who did not [17, 18].

These estimates of selective advantage raise an important point about natural selection. Most natural selection is not a life-or-death matter. Except in extreme circumstances, it is unlikely that people died because they were unable as adults to use the nourishment that milk provided. But if an individual in a tribe with cattle was only one or two percent more likely to survive and have children, because he or she could use lactose as an adult, this little boost was sufficient to ensure the rapid spread of the lactose tolerance allele.

And note that the tolerance alleles have not spread to everybody. In southern Europe, lactose tolerance is found in as little as thirty percent of the population. Are there counterbalancing selective forces that prevent the fixation of these useful alleles? The full story of lactose tolerance has yet to emerge, but it is likely to be complex. Even small shifts in allele frequencies can play an important role in evolutionary adaptation [19].

Breathing Thin Air

Food is important to our survival, but oxygen is even more essential. People who live in the high Himalayan valleys and on the Tibetan Plateau spend their entire lives at altitudes above 4,000 meters. At such altitudes the atmospheric pressure is about half the pressure at sea level, which means that there is only half as much oxygen in every breath of air to supply those friendly little mitochondrial mutualists that we carry in our cells.

Such a shortage of oxygen is most dangerous during pregnancy, because the fetus must obtain oxygen from its mother's circulation. Tibetan mothers have been selected to have an unusually high flow of blood to their placentas, carrying oxygen to their babies, during pregnancy. The result is that Tibetan babies are of normal size, while babies born to Han Chinese mothers living in Tibet are smaller than those who are born at sea level [20].

Our trips along the Everest trek route and into the high valleys of Bhutan were made easier by advice from U.C. San Diego's high-altitude physiologist John West and his Nepali colleague Buddha Basnyat. These colleagues also introduced us to people who have adapted to high altitude (Figure 72). And we met lowlanders from Nepal and India who are moving into the Himalayan regions in search of work. These migrants are also being subject to strong selective pressures for adaptation to high altitude.

The adjoining towns of Khumjung and Kunde occupy a wide rocky valley at an altitude of almost 4,000 meters on the way to the Everest base camp. A little

Figure 72 Population pressure and the lure of work have brought these lowland Nepali porters to high altitudes, where they carry burdens of almost 90 kilograms.

way up on one of the hills, the neat stone bungalow that houses Kunde Hospital has a view of the entire valley. Kunde's head doctor, Kama Thinba Sherpa, oversees a staff of seven. The hospital, which was started by the Hillary Foundation, serves about 4,500 people in the area, ninety percent of whom are Sherpas. Kama Sherpa's staff also treats the trekkers and climbers who pass through the area in increasing numbers each year. They must deal with everything from "Khumbu cough" (named after the valley that leads up towards Everest) to life-threatening effects of high altitude.

The physiological effects of extreme altitude can take three forms. In spite of its name, the least dangerous form is acute mountain sickness, a usually temporary condition marked by headache and loss of appetite. But this mild illness can progress into the much more dangerous high-altitude cerebral and pulmonary edemas.

These edemas have two major causes, both of which force fluids from the circulatory system into parts of the body where they do not belong. In the lungs, blood flow in the capillaries increases as the body tries to accommodate to an oxygen shortage. When the capillaries distend, fluid is forced out of them into the airways of the lungs. In other parts of the body, through mechanisms that are less well understood, liquid is forced from the circulatory system into the brain and other tissues, causing blinding headaches and retinal hemorrhages.

These effects are made worse if the victim is dehydrated or if the number of red blood cells is increasing as our bodies try to adapt to high altitude. The resulting thick, corpuscle-laden blood builds up pressure in the circulatory system to even more dangerous levels.

During our trek several tourists had to be airlifted out of the Sagarmatha Park area with symptoms of acute edema. Kama Sherpa told me that most of the victims he sees are young people who have pushed themselves too hard and have neglected to rehydrate. Liz and I, forewarned by John West, walked slowly and drank huge amounts of water and Gatorade that we reconstituted from powder. On the way up we spent a full day resting at 3,000 meters. And we stopped frequently on the trail, both to chat with local people and trekkers and to answer the insistent calls of nature that were triggered by our rehydration regime.

The sicknesses triggered by altitude are no problem for the local people. Kama Sherpa had not seen a case of high-altitude edema among the Sherpa population in his ten years at the clinic. He told me that if local people go to lower elevations to work and then return, they must readjust for a few days, but they never suffer from the symptoms that make high-altitude sickness so deadly.

Why are the Sherpas essentially immune to altitude sickness? Increases in the rate of blood circulation, which have been so valuable in aiding the survival of Sherpa and Tibetan babies, are part of the story. Another important factor is something quite counter-intuitive that happens—or rather fails to happen—to their blood.

Low-altitude people like Liz and I who blunder into this extreme environment compensate for the lack of oxygen by making more red blood cells. Sherpas and Tibetans do not. It is true that if they climb to altitudes higher than 4,000 meters they do make a few extra corpuscles, but the increase is not significant. A big boost in the rate of formation of red blood cells in response to high altitude is common among mammals, but this boost has almost disappeared in Sherpas and Tibetans [21].

Just as the Tharu people lost one of their α-hemoglobin genes, and the early agriculturalists lost the ability to regulate lactase secretion, the people of the Third Pole have lost most of their ability to make extra red blood cells.

Some of the genetic story of this adaptation is now emerging. Three international groups of scientists have scanned the entire genomes of Tibetans and Chinese lowlanders, looking for differences between them. Each group found that the Tibetans have a pattern of DNA polymorphisms on a piece of chromosome 2 that differs from the pattern seen in lowland Chinese [22–24].

Embedded in this region of Chromosome 2 is a gene called *EPAS1* that codes for a regulatory protein involved in the synthesis of hemoglobin. The *EPAS1* gene itself has not been altered in the Tibetans, so the DNA change that has been selected for must be a mutation in a regulatory region that lies somewhere near the gene. It will be years before we fully understand how this change has affected red blood cell production. But one intriguing surprise has already emerged.

The native American Quechua and Aymara of the high Andes have adapted to conditions that are almost as extreme as those faced by the Tibetans and the Sherpas. But, in contrast to the peoples of the Third Pole, they often suffer from a chronic form of mountain sickness that can shorten their lives [25].

When Liz and I visited Lake Titicaca's Isla del Sol in 2010, we met an Aymara woman and her endearing llama (Plate 16). Her village, Yumani, is on the highest ridge of the island, well above 4,000 meters. Unlike the Tibetans and Sherpas, when she moves to higher altitude the numbers of red cells in her blood increase. As a result she is at risk for altitude sickness.

Although the genomes of the Quechua and Aymara do show a number of differences from those of nearby lowlanders, there are no consistent genetic differences in their *EPAS1* region [26]. Is it possible that these people may not be as well adapted to high altitude as the people of the Third Pole? Perhaps. Again we see how important it is to have a diverse gene pool in the great game of evolutionary adaptation. The Quechua and Aymara are descendants of a small group of East Asians who migrated into the Americas about 14,000 years ago. This group brought with them a relatively restricted sample of genes, and the ability of the Andean peoples to adapt to high altitude may have suffered as a result.

In contrast, like the Tharu, the Tibetans and Sherpas are part of a web of migrant peoples with origins from Africa to Europe to East Asia. It is likely that the mutant alleles carried by these peoples have provided a much richer genetic resource for adaptation to new environments. There has also been a longer period of adaptation—modern humans probably colonized the Tibetan plateau between 25,000 and 30,000 years ago.

Some of these diverse pools of variation may be maintained by balancing processes, contributing to a genetic equivalent of the green equilibria that we saw in the world's rainforests. The Tharu carry more than one α-thalassemia allele. The Tibetans carry a variety of clusters of genetic markers around their

EPAS1 gene, suggesting that selection here too has been complicated and that more than one *EPAS1* regulatory allele has been selected for. Evidence is growing that a number of other genetic regions have increased in frequency without becoming fixed in the Tibetan population.

The Irish are the most extremely lactose-tolerant people of northern Europe, but they have achieved a frequency of the European lactase persistence allele of only 95 percent. Even when strong directional selection is acting on a population, instead of simply replacing an older allele with a newer one the selection usually moves the population from one complex equilibrium state to another [27].

Selective Sweeps

Whole-genome studies are now revealing signs of what appear to be thousands of selection-driven "sweeps" of pieces of DNA through our gene pool, marked by linked pieces of DNA that they have carried with them. These data suggest that human evolution over the last 80,000 years has been rapid [28]. But such selective sweeps can only be detected if genetic recombination has not had enough time to break up the linked pieces of DNA.

Now, it is possible to compare the differences between human and Neanderthal DNA sequences for signs of selection. This pushes the time-line for detecting possible selective sweeps further back, to as much as 500,000 years.

Richard Green and his coworkers have found over two hundred regions of the human genome that bear unusually little resemblance to the corresponding regions of the Neanderthal and chimpanzee genomes [29]. Selection has pushed many of these pieces of human DNA in a unique direction during our evolution. Other researchers have found that only a few of these selective sweeps would have been detected if the Neanderthal genome had not been available [30].

The reasons for most of these selective sweeps remain to be discovered. And, like the more recent lactose tolerance and *EPAS1* sweeps, these sweeps also tend to be incomplete. It is unusual to find a genetic marker that has swept through the entire human gene pool, so that is has become completely "fixed" since the time of the human–Neanderthal split [31]. Thus, natural selection has not pushed in the direction of a single "human" genome. Instead, different human lineages have been pushed in many different directions.

Now, as our world grows smaller and geographically separated people are beginning to meet and mate with each other, these independently selected alleles are coming together in a vast number of new combinations. The same thing must have happened when the pioneering bands of modern humans who first migrated out of Africa met and mated with the Neanderthals and Denisovans. Did these pioneers benefit from those introgressed genes?

Stanford immunologist Peter Parham and his colleagues have proposed such a benefit. They suspect that modern Eurasians may have inherited a valuable allelic form of a gene complex from Neanderthals. This complex specifies genes that are involved in our immune system [32].

At the moment the evidence for this contribution is highly circumstantial. The Neanderthals that have been sequenced did not actually carry this immune system allele, though they did have alleles that are closely related to it. But even if Parham's suggestion is not confirmed, his paper signals the start of a massive forensic DNA investigation that will eventually reveal the details of the Neanderthal contribution to our gene pool. We do not yet know how much of the Neanderthal genome has survived. It is possible that present-day Eurasians carry many different scattered parts of it. As the details emerge, it will be possible to determine whether natural selection played a role in the survival of these DNA sequences from our extinct relatives.

Natural Selection Continues

The trail from the airport at Lukla to the Everest base camp is lined with tea shops and little guest houses, built to serve the growing flood of climbers and trekkers. The Sherpas carry the visitors' equipment and supplies, and they have negotiated an agreement that no porter should carry more than thirty kilograms or 64 pounds. But every board and nail, every can of soda, and every roll of toilet paper that supply the teahouses must be carried up the trail by some person or pack animal. This far less romantic ecological niche is being filled by people from the foothills.

When we hiked across the tiny airstrip at Syangboche, near Khumjung, we encountered this adaptation in progress. Here, at 3,700 meters, young men from the lowlands staggered like laden ants under their burdens (Figure 72). They were headed for a teahouse construction site further up the trail, carrying loads that were too awkward for pack animals. Each of these porters was bent beneath two enormous sheets of construction plyboard, and each sheet weighed 44 kilograms, making a total load of almost two hundred pounds.

The porters who are best at handling this work will be the most successful at bringing money back to their villages, and this in turn will help them to marry and have children. It is often suggested that in the modern world, thanks to the comforts of civilization, any selective pressures acting on our species have vanished or been reduced to low levels. The people who have proposed this idea have not met these porters. As we observed their struggles, it was clear to us that natural selection continues to act on our species.

The people who live in this extreme part of the world are still caught up in evolutionary change. And we are far from the best organisms at adapting. Only the fittest mountaineers are able to stagger up the highest slopes of Mount Everest into the zone of death, where brain cells can be killed in large numbers by lack of oxygen. As they struggle, superbly adapted red-billed choughs flit past them in the impossibly thin air.

Although our adaptations are far from perfect, our complex genetic history has equipped us to colonize every part of the world. But we are only beginning to learn how this happened. Until recently we have blundered through the length and breadth of our planet in virtually complete ignorance both of our own history and of the consequences of our actions. We have drawn on our intelligence and our technology to exploit the world, but we have only just begun to explore our potential to heal the world as well. In the next chapter we will look at two of those early blunders. Because of different ecological and genetic circumstances, these stories have had different outcomes. What can we learn from them, as we wrestle with the task of restoring the world's green equilibria?

CHAPTER TEN

Learning From Our History

In this chapter we will examine two central events in the history of our species: the colonization of New Guinea that took place at the end of the first out-of-Africa migration between forty and fifty thousand years ago, and the arrival of the first humans in the Americas about 14,000 years ago.

Even with their limited technology, these migrants were responsible for profound, sometimes irreversible, changes in the green equilibria of the ecosystems that they invaded. But these events are not mere tales of the destruction brought about by our intelligent but thoughtless species. They are also stories of our resilience and adaptability.

It is well to remember that the people who took part in those invasions were caught up in the moment, and had no idea of the ecological and evolutionary consequences of their actions. Now, as we make similar profound changes to the entire planet, ignorance is no longer an excuse. Information is now available about exactly what we are doing, and exactly which ecological laws we are breaking.

An Explosion of Human Adaptation

The great migration of modern humans out of Africa had a diversity of unexpected results. As we saw in the last chapter, the ecosystems that the migrants encountered during their long journey shaped their cultures and even their genes.

Some of these ecosystems survived the human onslaught better than others. In this chapter we will use the ecological and evolutionary rules that we have explored in this book to help us understand this diversity of outcomes.

When the first bands of humans arrived in Australia, probably a little more than 40,000 years ago, the climate had already entered a long-term drying trend. The entire continent had become a vast tinder-box. The first migrants set fires that spread uncontrollably and destroyed many of the ancient acacia and evergreen forests.

Some of these dense forests were replaced by more open stands of eucalyptus trees, which made hunting easier. But in drier areas the forests were unable to recover and were replaced by deserts. These first generations of human colonists were forced to adapt to hardscrabble desert conditions that they were partly responsible for creating [2, 3].

Together with the fires, hunting had a devastating impact on Australia's marsupials. More than twenty genera of large marsupials disappeared when humans arrived. The victims included carnivores, grazers, and browsers [4].

Australia's green equilibria were so vulnerable that a few bands of Stone Age hunters were able to push them dramatically, and irreversibly, out of balance. This was primarily because of the dry climate, but the extinction of many

Figure 73 A   When Australian gold prospector Michael Leahy made the very first airplane flight over highland New Guinea's Wahgi Valley in 1932, he was astonished to find that the valley was being farmed intensively. Stone Age peoples, isolated for tens of thousands of years in New Guinea's highlands, had independently invented agricultural methods that used sophisticated systems of water channels and drainage. (From [1].)

Figure 73 B I took this aerial view of fields in the same valley in 2005. Aside from the buildings, the roads, and the greater variety of crops, the scene is remarkably unchanged from the one that greeted Michael Leahy in 1932. The agricultural practices that were invented by the highlanders have been so successful that they have persisted even after the outside world discovered these hidden valleys.

animals that occupied higher trophic levels, along with changes in the pattern of vegetation, magnified the effects of these human invaders.

At about the same time, the first humans arrived in New Guinea. They brought with them essentially the same technology that the Australian migrants possessed. But these people faced a different world, one that was wetter, richer, and less easily transformed. Even today, for each of New Guinea's ten million inhabitants, half a million cubic meters of rain fall on the island each year [5].

New Guinea's resilient green equilibria allowed these immigrants to radiate into the diverse cultural groups that now occupy its valleys and canyons. During that time they were able to take their stone-age technology in unique directions (Figure 73 A and B). Some remarkably advanced stone axe-heads have been found at a coastal site on the Huon Peninsula, not far from the present-day village of Yawan. Archeologists from Australia and Papua New Guinea have dated the axes to 40,000 years ago [6,7].

The makers of these axe heads deliberately shaped them to give them "waists," so that they could be firmly attached by bindings to wooden hafts (Figure 74). These waists are evidence of a startling technological advance. Such sophisticated axes were not being made anywhere else in the world at that time.

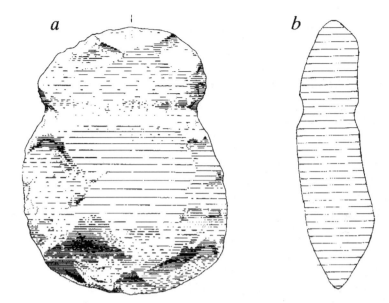

Figure 74 A "waisted" axe head, forty thousand years old, from the Huon peninsula still shows the indentation from the cord or vine that bound it to a wooden haft. (Part of Figure 3 from [6].)

With these hafted axes, the early migrants may have been able to clear the thick forests, chop down sago palms to extract their pulp, and build boats. But the axes were probably used mainly for warfare. Similar axes have been employed in conflicts between highland tribes down to the present time. And, through one of the ironies of history, it is exactly such murderous conflicts that have helped to generate the island's cultural diversity.

Cultural Diversity and New Guinea's Green Equilibria

The first people who ventured up into New Guinea's highlands from the coasts were able to escape the fevers and the limited food sources of the dense coastal forests. The highland weather was cool and spring-like for much of the year, and the cloud forests were filled with tree kangaroos and cuscuses. Delicious cassowaries were common, but dangerous to hunt—adult birds are able to disembowel incautious hunters with a sweep of their claws. The early migrants probably caught cassowaries as fledglings and raised them in cages, a practice that is still seen throughout the highlands today.

Soon, as the easiest food sources ran low, some of the immigrants began to exploit other features of their environment. The dwellers in these lush highland valleys—quite independently of the rest of the world—had the resources and the opportunity to invent agriculture [8, 9].

Excavations at the highland town of Kosipe, north of Port Moresby, show that highland people were gathering and cooking *Pandanus* seeds and yams as early as 41,000 years ago [10]. The excavations also uncovered waisted axe-heads, dating to 28,000 years [11].

The earliest firm evidence for slash-and-burn farming in the New Guinea highlands has been found beginning 7,000 years ago [12]. But farming is likely to have had much earlier origins, and it may have spread beyond New Guinea.

There is evidence that primitive agricultural techniques may have been used by the peoples of northernmost Australia, during the time when that continent and New Guinea were joined together. Scattered populations of bananas, taro, and yams grow unexpectedly in parts of Arnhem Land and the northern Queensland coast. These isolated populations suggest that there had been early cultivation of these important food plants [13].

Ten thousand years ago, northern Australia was a far lusher place than today's harsh environment of deserts, bush, and a few areas of open forest. Rivers that originated in New Guinea's mountains were able to flow unimpeded across what is now the Arafura Sea and the Torres Strait, into northern Queensland and the Northern Territories. New Guinea's Fly River flowed at least as far south as Australia's Gulf of Carpentaria [14]. But rising sea levels soon separated New Guinea from Australia, blocking the flows of both fresh water and new cultural ideas from the north. As conditions in northern Australia grew harsher, its tribes forgot their first hesitant forays into crude agriculture and fell back on hunting and gathering.

Australia's unforgiving climate stripped the technology of its human settlers down to the minimum that was essential for survival. In contrast, New Guinea's isolated highlands nurtured a remarkable cultural complexity and extreme behavioral divergence, as Liz and I discovered in 2009 when we visited the Baliem Valley in the western half of the island. There, and in other parts of New Guinea's highlands, we discovered how ecological green equilibria have helped to drive an explosion of cultural adaptations. Until recently, this cultural diversity was maintained by selective pressures that kept the human populations of the island in balance with their environment. But this balance, like other ecological and genetic equilibria, was maintained at a great cost.

The Baliem Valley is in West Papua, the half of New Guinea that currently belongs to Indonesia. In 1938, a plane from an American expedition made the first flight over the mountains that surround and isolate the valley. As the plane's pilot, Russell Rogers, flew across the valley's floor, he saw large areas of intensely cultivated farmland below. The scene was similar to the one that Michael Leahy had photographed five years earlier when he flew over the Wahgi Valley to the east.

When American explorer Richard Archbold led the first overland expedition to the Baliem Valley in 1939, he was driven back by fierce warriors. After his retreat, and because of the upheaval caused by World War Two, the valley's isolation continued unbroken for six more years.

Then, in 1945, a plane filled with U.S. troops crashed in steep mountains near the valley. Twenty-one of the twenty-four people aboard were killed. The U.S. Army swiftly mounted a spectacular rescue operation. Rescuers were parachuted to the site of the crash, and managed to bring the survivors down to the valley floor. There they enlisted the friendliest and least shy of the Stone Age tribespeople to help clear obstructions from a crude landing strip.

Once the strip was cleared, an army plane towed a glider from the coast to the valley, where it was released and floated to a landing. The rescuers and the rescued packed themselves into the glider. Then the tow plane swooped low, caught a loop in the glider's rope with a hook, and pulled it into the air. Crowds of tribespeople looked on in amazement as their new friends were yanked into the sky. Coverage by the world press of this spectacular operation brought the valley and its tribes decisively and permanently into the modern world.

The Baliem Valley's inaccessibility had been typical of the entire highlands. Travel was so difficult that the thousands of villages and clans that thrived in the highlands tended to develop their own customs and beliefs. Tribes even stayed separate if they lived in the same valley, because of incessant fighting between them.

Warfare was often ritualized, and the killing of a single warrior was sometimes enough to bring the hostilities to a halt. But more deadly conflicts were common. Kidnappings and cannibalism during these real wars guaranteed that vendettas would last for generations, further isolating the tribal groups [15].

When the first Australian gold-hunters arrived in Papua New Guinea's coastal settlements in the late nineteenth and early twentieth centuries, they often saw large numbers of bodies floating down the rivers. The bodies provided mute testimony that somewhere in those remote highlands people were slaughtering each other as they had for tens of thousands of years.

Among the highland tribes brief alliances based on intertribal marriages were common, but most of them soon broke up and the tribes went their separate ways. Intratribal quarrels could break out over a murder, over ownership of women, land, or pigs, or over many other triggers for mayhem [16].

The languages of these isolated tribes also diverged from each other, making the gulfs that separated the different cultures more unbridgeable.[1] The original migrants' languages multiplied into more than a thousand, a number that is likely to be an underestimate because remote parts of West Papua have yet to be surveyed.

At the time of the Baliem Valley's discovery, most of its inhabitants were members of the Dani group of tribes. Their lives were a mixture of ancient stone-age customs and newer agricultural techniques, many of which they had invented to take advantage of the new foods that had percolated in from the coast.

Much of the valley has changed during the single long human lifetime since the first planes flew over it. But some aspects of the lives of its agriculturalists and hunter-gatherers have remained the same.

Tim Denham of Australia's Monash University and other archeologists have found that the earliest rather equivocal traces of true agriculture in the valley date back 7,000 years. By 4,000 years ago, elaborate systems of drainage ditches were being constructed. But a kind of bridge between simple gathering and true agriculture may have begun much earlier, just as it may have done in northern Australia. Pollen records show that the numbers of native banana and *pandanus* plants began to increase in the highlands starting about 40,000 years ago—though it is possible that climate change, rather than deliberate cultivation, was responsible for their spread [11].

Over thousands of years the valley's isolated inhabitants learned how to build small raised beds of earth in order to raise sweet potatoes, taro, and bananas, and to cut drainage ditches that helped to drain water away from their precious plants. Traces of over forty different species of edible plant have been found during the excavations of these early gardens. A complex set of cultural interactions developed, stimulated by trade and ownership of pigs, which were first introduced from the coast at least five thousand years ago (Figure 75) [18].

Much of this tradition remains intact in the valley. We visited the beautiful stockaded village of Wasilimo, where the village men rather ineptly slaughtered the pig that we had brought. They then butchered it using razor-sharp bamboo knives. The women filled a firepit with hot stones and damp grass, and layered it

Figure 75 Two Dani pig herders rest on a rock outcrop overlooking the lush Baliem Valley in West Papua. In daily life the tribesmen wear penis gourds, string bags called bilums, and little else.

with pieces of sweet potatoes and yams. They carried the stones to the firepit using split saplings as giant tweezers. The men layered the bloody pieces of pig above the tubers. Finally, the women bound the whole heap expertly with vines, squeezing it together so that steam from the heated grass would permeate the entire structure.

The cooking process was remarkably efficient. In less than two hours the food was ready to eat. The men gorged on the pig and handed out small pieces to the women and children.

In spite of appearances, the village is changing. When we talked to the women through our interpreter, we discovered one change that has had huge consequences for their lives.

All the first joints of the fingers of the older women in the village had been amputated (Figure 76). Traditionally, whenever an important man in the village died, young girls were chosen to express the village's grief. They were struck on the elbow with a club, and the tip of one of their fingers was amputated during the momentary mixture of agony and numbing that followed. They showed us crude stone knives, distressingly blunt, that had been used in these operations.

189

Figure 76 My wife Liz compares her intact fingers to the shortened fingers of the Wasilimo women. The fingers of the young girls to the left are intact, showing how thousands of years of tradition are now beginning to change.

How and why this mutilation ceremony arose is unclear, but it is widespread in New Guinea and probably very old. Micheal Leahy observed it in tribes of the Bismarck Range in the 1930s.

Such cruel amputations are now rare, and throughout the island other behaviors are changing as the peoples of the highlands begin to encounter the modern world. Some of the cultural diversity of the highland communities will be lost—undoubtedly a good thing in the case of finger amputations and the grim traditions of murder, cannibalism, and warfare. But some of their diversity, a gift from New Guinea's green equilibria, will be retained.

Human Conflict and Green Equilibria

The first migrants who arrived in New Guinea depended on the animals and plants of their forests and valleys to support high populations and an explosion of cultural diversity. They augmented these resources with a growing reliance on agriculture and animal husbandry. But they seem not to have overexploited the capacity of their world—at least not until the recent flood of new technologies from the outside.

Different ecosystems on the island were preserved for different reasons. Diseases, food shortages, and poor food quality limited the sizes of coastal and lowland tribes, and may have helped to preserve the coastal rainforests until recently. Now these accessible coastal regions of the island are changing rapidly. The coastal flatlands that lie between Lae and Madang on Papua New Guinea's northern coast, once heavily forested, have been utterly transformed by farming and ranching. When I drove through those sun-blasted plains, I was greeted by a Texas-like landscape of dust and scattered, stunted trees.

In contrast, New Guinea's highland environment has lucked out. The Australian Aborigines were easily able to burn off their new lands, but the tribes who moved up into New Guinea's cloud forests found themselves in a soaking wet environment that was difficult to burn. This was lucky, because if they had managed to destroy the forests their opportunity to invent agriculture might have disappeared along with their rich and complex environment.

For tens of thousands of years the peoples of New Guinea were directly influenced by the green equilibria of their environment. The result was a vast experiment in the generation of divergent social systems, the real-world equivalent of the spaceship-borne transplantation of colonists that I speculated about in Chapter Six.

The results of this natural experiment are displayed each year in the great cultural festivities called sing-sings. The first sing-sings were organized by missionaries in the 1950s, in order to channel between-tribe aggression in more productive ways. The sing-sings drew on hundreds of far smaller traditional tribal festivals, and brought together many isolated tribes for the first time.

Liz and I attended two different sing-sings in 2005, one at Mount Hagen in the Wahgi Valley in the Eastern Highlands and one that drew its participants from fifty coastal villages near the town of Tufi on the island's eastern coast. The sing-sings were joyful celebrations that combined simulated warfare and displays of music and costume. Especially at the Mount Hagen highland sing-sing, the cultural diversity on display was astonishing.

Plate 17 shows a small sampling of the highland cultural diversity, mostly from the Wahgi Valley area, that was on display at the Mount Hagen sing-sing. Clockwise from upper right: (1) A woman and a boy from the lower Chimbu re-enact a mourning ceremony. (2) Huli wigmen spend months growing their own hair into wigs that can act as cushioned helmets in battle and protect them from those hafted stone battle-axes. (3) Preparations for the sing-sing include a new role for the bottles of white-out that some of us remember from the ancient days of typewriters. (4) Until recently, their Asaro mudmen used their wicker-and-mud masks to terrify neighboring tribes into thinking that they were ghosts. (5) Chimbu skeleton men used a similar scare tactic to threaten their foes.

There are deep connections between the cultural diversity displayed at this highland sing-sing and the evolutionary history of New Guinea's ecosystems. Some of the participants' ornaments are made from feathers of the birds of paradise, and some of their dances are modeled after the birds' mating dances.

Plate 18 shows a Tapuka villager from the central highland Wahgi Valley. His nose is pierced by tail feathers of the King of Saxony bird of paradise, and his headdress is decorated by feathers of the blue and superb birds of paradise.

You will recall that the birds' plumage and dances had themselves evolved in the relatively predator-free highland and north coast forests, an evolution that was powered by unleashed sexual selection (Chapter Five). The displays of the different tribes have also diverged from each other, in part because of strong *intrasexual* and *intersexual* selection pressures. The male warrior displays have been driven by intrasexual (within the same sex) competition and by the need to drive off warriors of nearby tribes. Intersexual selection, in which both male and female mate choice plays a role, has resulted in the tribes' colorful costumes, spectacular dances and musical traditions.

The tribal differences are of course cultural rather than genetic. Gene flow among the tribes, a consequence of the many raids in which children were kidnapped, has certainly slowed any potential separation of gene pools. But there is no biological reason why, if New Guinea had stayed isolated for hundreds of thousands of years, these cultural differences and the accumulation of isolating mechanisms might not have eventually led to speciation (Chapter Six). You will recall that the diversity of the African environment had earlier contributed to the evolution of what appears to have been several hominan species.

Ironically, the diverse cultures of the highlanders may have helped to preserve their environment. Because these highly territorial tribes were isolated by topography and a tradition of warfare, they could not easily be conquered by nearby tribes. Until the Christian missionaries began to make the tribes' local sing-sings more inclusive, the members of each tribe were distrustful of their immediate neighbors.[2]

The tribespeople were not able to come together in large enough numbers, and over a large enough area, to overcome their differences and to allow an overarching cultural pattern to emerge. Even though some of the tribes practiced agriculture, they never established hierarchal agriculture-based civilizations of the type that intermittently thrived and then crashed in the Americas—and that flourished in the Middle East but eventually caused the destruction of many of that region's ecosystems [19].

What emerged in New Guinea was a kind of frequency-dependent cultural balance, in which any tribe that became too large or powerful was soon driven down to lower numbers and forced to retreat into a smaller territory by unrelenting warfare with the tribes that surrounded them. Of course, such a simple model of negative frequency-dependence glosses over many of the complex interactions among these isolated tribes. But conflict-based frequency-dependence is likely to have contributed to the preservation of both tribal cultural diversity and—serendipitously—the green equilibria of the highlanders' natural world.

In this book we have emphasized repeatedly the need to maintain the world's green equilibria. But we must remember that the maintenance of all of the green equilibria in ecosystems comes at a high cost to their animals and plants. Predation, disease, and starvation all play an essential role in the preservation of ecological diversity.

Because of its frequency-dependent effects, the warfare among the highland tribes has also played a role in preserving their environment. But this preservation has been accompanied by a terrible human cost. During all the time since the highlands were settled, a huge span of years that stretches for ten times as long as recorded history, the bodies of countless victims of these conflicts floated down New Guinea's rivers from the highlands to the sea.

In the modern world, the merciless natural selection of warfare is no longer an option for the maintenance of tribal diversity. Without these pressures, the extremes of cultural differences will inevitably be reduced. But there are compensations. After their long isolation, the people of New Guinea have become part of the world community. Their health and education are improving, and intertribal wars are becoming rare. Now, like the rest of us, they will be faced with the necessity of coming to terms with the sustainability of our planet.

The tribes of New Guinea's highlands have adapted to their world in many remarkable ways. Their rich environment, along with their rich genetic heritage, gave them the resources to invent agriculture, something that was denied by an accident of climate to the genetically similar peoples of arid Australia to the south. As we look at this long history of invention and adaptation, who can doubt that, like the rest of us, these peoples at the far end of humanity's greatest migration possess all the capabilities that are needed to deal with the current world's problems?

The Invasion of the Americas

There are parts of the Americas where green equilibria are still largely intact. But, starting right from the moment when humans first arrived on these vast continents, most of their ecosystems have suffered an ecological onslaught that has been intense and transformative. And as in New Guinea, the human invasions themselves were shaped by the laws of ecology and evolution.

The green equilibria that remain can still provide magical moments for the lucky observer. I experienced such a moment when I encountered an ocelot (*Leopardis pardalis*) on the bank of Guyana's Mapari Creek during one of our night drifts down the creek. Seconds after I took his picture (Figure 77), our boat crashed into a tree that had fallen across the creek. The ocelot, sensibly, vanished at the sound of the impact. I did manage to get an additional picture of his rear end as he was disappearing, which provided emphatic proof that he was a male.

Figure 77 An ocelot on Mapari Creek provides a snapshot of a world before it was changed by the arrival of humans in the Americas.

Jaguars, leopards, and ocelots still thrive in the upper reaches of Guyana's rivers. On his 2004 expedition to the upper Rewa with a group of BBC photographers, Ashley Holland had spotted ten jaguars. Pressures on such flagship animals are growing, however—just before our Mapari trip I learned that gold miners had shot five jaguars not far from the boundaries of Kaieteur National Park.

These human pressures are relatively recent, and they add to the strong selective pressures that such animals already face from their natural environment. You can see from this ocelot's face that he has suffered traumatic encounters— almost certainly with other males. His left eye is damaged and his nose is scarred and twisted. Among ocelots, intrasexual selection can take the form of fierce battles. These fights are doubly hard on the losers—not only do their injuries lower their chance of survival, but they also miss an opportunity to reproduce.

The ancestors of ocelots and other big cats arrived in South America during the Great American Interchange. If we count only the larger animals, the flood of new arrivals from North America included mammoths, mastodons and at least two other relatives of the elephants, along with predatory cats, bears, dire wolves, tapirs, two diverse groups of horses, and a swarm of species of camels, deer, and wild pigs [20].

Perhaps ten or eleven thousand years ago, the first humans arrived on the Guiana Shield. The shield's ecosystem was so rich, and these first migrants were so few, that their impact was small. But they were part of a great human migration that had far larger effects elsewhere in the Americas.

A few traces of these first settlers survive. One is a petroglyph-covered rock in the middle of the Kurupukari rapids of the Essequibo River (Figure 78). The age of the carvings is uncertain, because any datable artifacts that might have been left behind by the people who carved them have long since been washed away in the rushing waters. They are still regarded with superstitious awe by some of the villagers. It is considered bad luck to look at them, much less to photograph them—as of course I promptly did, thereby tempting fate.

We do not know how these people lived, what happened to them, or whether they had any descendants. We do know that, long after the petroglyphs were carved, members of the peaceful Arawak and the more warlike Carib tribes migrated north from the Amazon basin into this region, making their way through gaps in the *tepuis* to the flat coastal lowlands of the Guiana Shield.

Even then these lowlands continued to be a bit of a backwater. The people who arrived were still few in number, and most of them settled on the coasts.

Figure 78 These mysterious petroglyphs, which show dancing or gesticulating human figures, are revealed only when the rapids of Guyana's Essequibo River drop during the dry season. We know nothing about the age of these elegant carvings or about the people who made them. But similar incised figures have been found in the Brazilian Amazon, and have been tentatively dated to about eleven thousand years ago [21].

This accident of history has helped to preserve some of the diversity of the ecosystems in Guyana's interior.

Because of its remoteness, Guyana escaped the immense changes that were taking place elsewhere. Evidence is growing that long before recorded history began, and long before the current tsunami of human population growth hit the planet, humans succeeded in altering American ecosystems as dramatically as—and much more rapidly than—all but the most violent existential crises of the more distant past. But the evidence that connects these first Americans to these changes, like the evidence that an author scatters through a well-plotted murder mystery, is circumstantial. Exactly what happened, and why?

Until a geological moment ago, many animals that were thriving in the Americas left plentiful traces in the fossil record. Then most of them suddenly went extinct. Carbon-14 dates, some of which can be obtained directly from the bones of the animals themselves, show that most of these species vanished

PLATE 11 Some members of the Fish Market eelgrass community at Anilao.

PLATE 12 Scenes showing the abundance of Brazil's Pantanal wetland.

PLATE 13 Santa Cruz Island today.

PLATE 14 Plot workers are dwarfed by *Ficus archiboldiana* in northern New Guinea.

PLATE 15 A peach-faced lovebird (right) nests in a colony of social weaver birds (left).

PLATE 16 This Aymara woman has adapted to high altitude in different ways from the people of Tibet.

PLATE 17 Cultural diversity at the Mount Hagen sing-sing.

PLATE 18 Bird of paradise feathers decorate a highlander.

PLATE 19 Rhododendrons in Nepal's Himalayas.

approximately 13,000 years ago. And there is abundant evidence that their populations were thriving right up to the moment that they disappeared [22].

In North America, this wave of extinctions wiped out about 35 genera of large mammals, including mammoths, mastodons, saber tooth cats, camels, horses, giant sloths, and giant beavers [20]. A similar wave of extinctions in South America, beginning roughly a thousand years later, caused the disappearance of more than fifty genera of large mammals. Among them were many relatives of the camels, many endemic species of horse, placental saber tooth cats and other giant cats, giant ground sloths, and giant armadillos. Although the North American extinctions have received the most attention, the losses in South America were actually more extensive.

These extinctions were almost exactly coincident with the first wave of human immigrants from eastern Asia. Starting about 14,000 years ago people moved rapidly south from Alaska down the length of both continents. Paleontologist Paul Martin proposed that human hunting was directly responsible for these and other extinctions around the world. He termed these events "Prehistoric Overkill" [23].

There is a good deal of hard evidence for Martin's hypothesis in other parts of the world. When humans arrived on Madagascar, the giant flightless elephant birds eventually disappeared, probably because people stole and ate so many of their eggs. And there are historical records that clearly implicate New Zealand's Maoris in the extinction of the islands' giant flightless moas.[3]

Charcoal traces from lake bottoms show that when the Maoris arrived on the wet and heavily forested South Island of New Zealand 750 years ago, they still somehow managed to burn the island from one end to the other [25]. It is easy to see how Maori hunters, aided by dogs and fire, could have chased down the last desperate moas in even their remotest refuges.

But North and South America are huge compared with these oceanic islands. How could small bands of Stone Age hunters, dwarfed as they must have been by the vast continental landscapes, possibly have driven so many different kinds of formidable and cunning animals to extinction?

Circumstantial evidence of human involvement in these extinctions is nonetheless growing. Owen Davis of the University of Arizona and his colleagues ingeniously tracked the abundances of large grazing animals at several North American sites by counting the spores of a fungus, *Sporormiella*, in soil samples [26]. This fungus grows on herbivore droppings, and the numbers of its spores fell dramatically approximately 12,900 years ago.

Nonetheless, direct evidence of human involvement in this decline is sparse. The oldest apparent kill site yet discovered in the Americas, on the Olympic Peninsula in the state of Washington, has been dated to about 13,900 years ago. A long, sharpened fragment of bone was found firmly buried in one of the ribs of a mastodon at the site. This was not a natural injury. DNA sequencing and protein analysis showed that the bone point was also from a mastodon. It must have been part of a hunting weapon [27].

The forensic evidence from this site is especially clear, but only a few other sites have been found at which the bones of animals show signs of butchering or of wounds caused by human weapons. Horses were butchered about 11,000 years ago at several sites along the Alberta–Montana border. And two incontrovertible sites where mammoths had been killed, dating from about the same period, have been found in southern Arizona. About 12,800 years ago, a herd of bison was driven up an arroyo in Oklahoma and apparently slaughtered with bows and arrows [28].

In South America there are some scatterings of mammoth, giant sloth, and horse bones in central Chile that show signs of butchering. At one Chilean site, Taima Taima, an arrowhead was found lying inside the skeleton of a partially butchered young mastodon [29].

Opponents of the overkill hypothesis have seized on the rarity of such sites. Proponents have countered this argument by pointing out that kill sites are likely to be rare, because the animal remains at them would probably have rotted or been scattered by other animals and because most mortality in even heavily-hunted populations is likely to be from natural causes.

In other parts of the world, ancient kill sites are also sparse. Such sites that date to the first arrival of humans have not been found in Australia. But Australia suffered a wave of extinctions of large marsupials at that time, and evidence for a human role is growing.

In Europe and the Middle East, a mere dozen sites have been found at which mammoths were possibly killed by Stone Age hunters. Unlike the sites in the Americas, which are closely grouped in time, these Eurasian sites span more than 200,000 years. Arrowheads buried in bone were found at only two of them [30]. At most of the sites the animals might simply have been butchered after they died. The fact that a comparable number of sites have been found in North and South America suggests that hunting in the New World, which took place over a much briefer span of time, might actually have been more intensive.

What other explanations could there be for this swift wave of extinctions? Donald Grayson of the University of Washington, Russell Graham of the Illinois

State Museum, and their colleagues have implicated climate change [31]. And indeed, the North American extinctions did take place at the same time as the rapid warming that marked the end of the last ice age.

But North American climate change does not explain South American extinctions. Southern South America actually began to warm up earlier than northern North America, even though the southern mammal extinctions took place four hundred to a thousand years later than those in North America [22]. And there are other discrepancies that call into question the climate change hypothesis. More than a hundred thousand years ago, long before the arrival of humans, there had been an even more rapid warming period. Yet this environmental shift did not trigger a comparable burst of extinction. [20]

It is of course possible that human impacts and climate change might have interacted synergistically. Reductions in the numbers of large animals, as their populations adapted to changes in climate and vegetation, could have been turned into precipitous declines by human hunting and fires. Together, these factors might have pushed many animal populations to the point of no return [32].

What about a rock from outer space? Richard Firestone of Berkeley's Lawrence National Laboratory and his colleagues proposed that a comet or asteroid impact, perhaps near Hudson's Bay in Canada, triggered the extinctions [33]. Alas, the evidence for such an impact, slender to begin with, has evaporated on closer inspection—and, once again, an event centered in North America could not explain the later wave of extinctions in South America.

Other explanations, such as the spread of diseases or the loss of food sources, also seem likely to be minor contributing factors, because the extinctions across the Americas involved many species and many different ecosystems.

In short, evidence for a central role of clever humans in these extinctions continues to accumulate.

Population Pressure and the Fate of Ecosystems

Population explosions, with their accompanying environmental overexploitation, have taken place repeatedly during the history of our species. But, because we are able to learn from our experiences, such population overshoots have often been followed by a period of adjustment.

The first human immigrants to North America had to be tough and resourceful. Forty thousand years ago, during a slight warming between ice age peaks, their ancestors had trekked east across the land bridge of Beringia that connected eastern Siberia to Alaska. There they found themselves in a daunting Arctic wilderness, covered with glaciers except for tiny bits of coastal plain.[4]

Mitochondrial chromosomes provide clues to the extreme conditions that must have been faced by these early colonists. Recall from Chapter Six that the descendents of the first modern human migrants out of Africa carried only a small fraction of the mitochondrial diversity that people still possess on that continent. Similarly, Native Americans of today carry a small subset of the mitochondrial diversity of their East Asian ancestors. It was during the long isolation of those first Alaskan populations that the mitochondrial diversity was lost, as these tiny bands were repeatedly driven to the brink of extinction by severe weather and food shortages [34].

Fourteen thousand years ago, as the world warmed, some of those first Alaskans were finally able to move south. They may have traveled along inland routes newly freed from the ice, but evidence is growing that they used their fragile boats to thread the coastal waters [35]. In a little more than a thousand years, some of their descendants had penetrated the furthest reaches of South America.

The ancestors of these first Americans had already honed their survival skills on the steppes of central and eastern Asia. There they had driven Eurasian wild horses, woolly mammoths, and woolly rhinoceroses to extinction [36]. When the retreat of the Alaskan glaciers unleashed these expert hunters on two brand-new continents, it is not surprising that their impact was so devastating.

It is probable that the availability of such huge resources resulted in rapid and unsustainable population growth. But we cannot be sure, because the effects of any early population explosions have been lost to history. In fact, we don't even know how many people were living in the Americas, including those who were part of their great civilizations, when Columbus arrived (Figure 79). Our estimates of the population of the pre-Columbian Americas range wildly, from ten million to a hundred million [37, 38].

Guesses about the population trajectories of the first people to arrive in the Americas are beset with even more uncertainty. But it is obvious that, as these bands of tough hunters traversed the length and breadth of the two vast continents, they were greeted with apparently limitless food and space. Their technologies would have had ripple effects that they could not have foreseen.

For cultural reasons, and driven by the imperatives of sexual selection, the human hunters were likely to have paid much attention to the most spectacular

Figure 79 Machu Picchu has grabbed all the attention, but there were many civilizations in addition to that of the Incas that occupied the Andean valleys before the arrival of Europeans. The cyclopean walls of Kuelap, a hilltop fortress in northern Peru, were built by the Chachapoya people 800 years before Machu Picchu. Kuelap was the largest pre-Columbian structure in the Andes, requiring more than twice as much stone as Egypt's Great Pyramid. In spite of the apparent invulnerability of this fortress, the Incas conquered it and subjugated the Chachapoya before the arrival of Europeans.

of the top predators. The most skilled and daring hunters returned to their villages with the dripping heads and skins of dire wolves, ground sloths, saber tooths, and especially the giant American lions *Panthera atrox*, the largest lions of which we have record. These proofs of their bravery must have paid off amply in wives and wealth.

The first Americans also probably set great uncontrolled fires, as their descendants were still doing when the first Spanish explorers arrived in California in the sixteenth century [39]. The fires would have altered the mix of vegetation that provided grazing, browsing, and protective shelter for many animals. The mammoths and mastodons, with their immense appetites, would have suffered disproportionately.

The effects of these changes were rapidly propagated down the trophic levels. As top predators were removed, grazers and browsers multiplied and began to outstrip their resources. Saber tooth cats and other large predators required great herds of prey animals from which they could cull the young and the sick. As the herds of grazers and browsers exploded and died back, becoming more scattered, the populations of their predators also crashed.

But what about the vast herds of bison on the North American Great Plains that greeted European explorers? In the eighteenth century they probably constituted the largest population of grazing animals on the planet. Surely their immense numbers would seem to argue against such overexploitation by the hunters of the plains. But the bison herds may have been a relatively recent phenomenon, and a symptom of ecological imbalance. There is no mention of these animals in the accounts of sixteenth century Spanish explorers of the southern plains, even though they should have been plentiful. Archeological sites show that the Plains Indians had exploited bison heavily by driving whole herds over cliffs, even before they began to use the horses that had been reintroduced by the Conquistadors. It has been suggested that the plains tribes extended the grasslands because of the fires that they set. Then, when the tribes were decimated by European diseases, these grasslands were available to support a population explosion of bison [37].

As the populations of American animals plummeted, so did those of the hunter-gatherers. These population crashes, probably involving both starvation and innumerable wars over resources, would have happened too late to save the roughly one third of North and South American mammal species that disappeared.

Because these tribes had no idea of the consequences of their overexploitation, population booms and crashes were unavoidable. Such overshoots and crashes happened elsewhere. We know from Icelandic written records, starting from the time that Iceland was first colonized in the ninth century, that its population overshot its resources and crashed at least twice [40].

The population growth rate of the first American hunter-gatherers need not have been especially rapid in order to produce unsustainable numbers. Suppose that a thousand people had made their way south through the paths that had opened up through the Alaskan glaciers. Suppose further that this population, which was able to draw on unlimited resources, grew at two percent a year. This rate may seem high, but it is probably conservative. The Egyptian population is growing at two percent each year at the present time.

At this growth rate, the early migrants would have doubled in numbers every thirty-five years and would have reached a million in 350 years. But this breakneck pace could not have continued. If their numbers had grown at the same compounded rate for another 350 years, they would have reached a billion. This is an obvious impossibility for pre-agricultural peoples.

When the populations of these early Americans collapsed, the survivors were faced with the consequences of their overexploitation. They and their descendants were forced to adapt. Eventually, independently of the rest of the world, some of these tribes invented agriculture and began to practice artificial selection on food plants with astonishing success. Over thousands of years, in some of the most remarkable accomplishments of preindustrial peoples anywhere in the world, these proto-agriculturalists produced highly selected strains of corn, potatoes, beans, tomatoes, squash, and many other crops. It is one of history's ironies that these plants, especially corn, have played a major role in fueling our recent worldwide population (and obesity) explosion.

The peoples of the Americas were as resourceful and clever at shaping their environment, and adapting to it, as the peoples of the New Guinea highlands. Thanks to the Europeans and their diseases, we will never know whether they would eventually have reached a new ecological equilibrium. But we are now able to detect the results of their repeated overexploitation of their resources.

In this book we have repeatedly seen how our history, our genetics, and the ecosystems that we have invaded have all shaped our species and our world. At the same time, our unique history has equipped us to undo the damage that we have done. In the next chapter we will explore how our brains have evolved in ways that allow us to meet this challenge. We will also discover that a complex balance of selective forces has added a kind of intellectual green equilibrium, a multidimensional layer of abilities, to our species as a whole. In short, we have no excuses—our brains are indeed collectively capable of overcoming our past mistakes.

CHAPTER ELEVEN

Green Equilibria and the Origin of Our Pretty Good Brains

U nlike most other animals, humans have the mental equipment to extract meaning from the world (Figure 80). Our brains have grown disproportionately compared with our bodies in the course of our recent rapid evolution [1]. We have already seen how some of this change may have been driven by natural selection. In this chapter we will explore how the evolution of our brains has involved much more than a simple increase in size. Just as with the other types of human adaptation that we examined in Chapter Nine, some of the evolution of our brains has been shaped by balances between opposing frequency-dependent selective forces. These forces, like the forces that have helped to establish the green equilibria of the world's ecosystems, have added to the richness and diversity of our intellectual abilities.

Controversy continues about how and why it was the genus *Homo*, rather than the other lineages of African hominans, that began to evolve larger brains. Charles Darwin (as usual) was the first to suggest convincing mechanisms. In *The Descent of Man and Selection in Relation to Sex* (1871) he pointed out that sexual selection has driven the pronounced physical differences between the human sexes. Perhaps, he thought, females also tended to choose smarter males. (He did not, alas, explore the possibility that males might have been attracted to smarter females.)

Darwin's explanation begs the question of why, in our species, sexual selection might have made such a large contribution to brain function. Many other animal species are subject to strong sexual selection, but have not become any more intelligent as a result. But in the same book Darwin took a further step in what I suspect was the right direction. He pointed out that any tribe of our ancestors must have included a small number of men who occasionally invented new kinds of communication or technology. On average these men, their relatives, or other members of the tribe who were smart enough to take advantage of the

Figure 80   These complex Aboriginal petroglyphs, on an outcropping of Central Australia's Uluru (Ayers Rock), show that modern humans have been able to distill meaning from their world even under the most challenging conditions. In this chapter we will look at how our brains have acquired this and many other abilities.

new advances, would be more successful at reproducing than those who were bewildered by the changes.

Darwin pointed out that the ability to invent things, or at least the ability to take advantage of inventions, would have been passed on to subsequent generations and would have led to further advances. In common with almost all of his contemporaries he ignored the immense contributions of women to human cultures and technology. But it is obvious that the mechanism he suggested becomes even more powerful when the power of natural selection for inventiveness, acting on both sexes, is added to the power of sexual selection.

Darwin's suggestion can be thought of as a feedback loop between the growing complexity of our physical and social environment and the ability to make it even more complex. We now know, as Darwin did not, that our changing gene pool has been an essential part of this loop, which involves our brains, our bodies, our environment, and our genes [2].

I would like to suggest that this Darwinian feedback loop can be traced back at least to the time of *Ardipithecus*, our ancestors from 4.4 million years ago who traded fierce canine teeth for friendlier smiles as they began to live together in groups. The communal environment of these proto-tribes might have provided opportunities for new technologies and new means of communication to arise.

For most of the time since *Ardipithecus,* technology has changed slowly. Our ancestors used stone tools that altered very slowly for the better part of two million years [3]. It is likely that the feedback loop would have been able to keep our genes synchronized with social and technological change through much of this time.

Eventually, however, the rate at which we were changing our environment outstripped our ability to evolve to keep up with it. By that point, perhaps well before the invention of agriculture, we had become clever enough to invent so many things that our gene pools could not keep up.

In 1998, I suggested that this feedback loop has been powerful enough to provide all of us with pretty good brains [4]. Our intellectual and cultural evolution has equipped us to understand the world in ways that are not available to any other organism on the planet. Now we are faced with the necessity of both re-establishing the green equilibria of our planet and adapting ourselves to them. We must ask: are our brains up to the task? Are our cognitive abilities, which got us into this ecological mess, good enough to get us out?

The simple answer is yes. The brains of all of us constitute a resource that we have barely begun to tap.

The precise reasons why our brains have become such an important resource require a more complicated answer. We can begin with the fact that our brains have been shaped by evolution to be robust, highly redundant, problem-solving devices. This has happened even though increases in brain size have come at a severe energetic cost.

Human brain tissue consumes energy at about ten times the rate of the rest of our bodies. Leslie Aiello of the University of London and her coworkers have proposed that increases in brain size must have coevolved with the ability to extract energy from food with greater efficiency. They gathered evidence that the digestive systems in animals with large brains have become smaller, as their diets have become richer and better able to support their expensive brains [5].

Alas, the evidence that was used to support this "expensive tissue" hypothesis has faded away with the gathering of complete anatomical information from more species. The pattern that has now emerged is that species with larger brains tend to have smaller fat reserves, not smaller digestive systems. There are, it seems, two alternative strategies to surviving bad times—you can be clever, skinny, and active, which helps you to find extra food, or you can store enough fat to be able to sit stupidly about and wait until the good times roll again [6, 7].

Big brains are costly even if the cells that make them up are not doing anything. Even when idle, their cells have to maintain a substantial charge differential across their cell membranes.

As a result of this charge difference, which allows the cells to stay at a high level of physiological readiness so that they can react rapidly to stimuli, even a resting brain gobbles up energy. And the more complex the neural tissue, the more demanding it becomes. For example, the eyes of different species of fly have neural photoreceptors of different sizes. Larger photoreceptor cells can transmit more information than smaller ones, but it has been shown that even when they are inactive they use more energy [8].

Although brains are metabolically expensive, they provide astonishing capabilities that can be essential to survival. Bats are able to integrate echolocation information literally on the fly, enabling them to thread through tree branches at high speed in pitch darkness [9]. Elephants and apes can apply "insight learning" to invent ways to reach food. And Alex, the famous African gray parrot, could employ language creatively, add numbers to reach totals as high as eight, and even flirt with the concept of zero [10].

In view of these and the many other remarkable accomplishments of the brains of other animals, we are forced to conclude that there are no properties of human brains that are truly unique. So why, given the huge metabolic cost of neural tissue and the fact that smaller brains can still do amazing things, did the brains of our ancestors begin to increase rapidly in size about two and a half million years ago?

Clearly, we are smarter than other animals. And our brains have evolved to enable us to do a wider variety of things. But, unlike other animals, the evolution of our brains has taken place in concert with recent changes in an unusually wide variety of other characteristics: changes in our postures, our hands, our feet, and our voiceboxes, along with the growing complexities of our social structure and technology. It is the sum of these changes, driven by Darwin's feedback loop, that has driven the evolution of our brains.

All these changes have happened so quickly that there might not have been much time to refine the process. If a simple increase in the size of the brain was enough to increase the number of possible connections between brain cells and provide our ancestors with an increased ability to adapt to their increasingly complex environment, this would have been selected for even if the increase also had some negative effects. The amazing brains of bats have been selected to be small and light, but our own ancestors' brains may not have been under such severe constraints.

As we develop, our brains grow in size at just the same time as we are first being exposed to the world around us. The size of the birth canal keeps the heads of human newborns small, so that our brains at birth are constrained to be about the same size as those of newborn chimpanzees. After birth the brains of chimpanzees increase in size only slightly. But as we humans grow to adults our brains increase in size more than threefold. This period of growth provides plenty of opportunity for interactions between our developing brains and the world that they are perceiving and reacting to.

This increase in size is not due to an increase in the numbers of *neurons*, the cells in our brains that transmit nerve signals. Instead, each of our neurons sprouts unusually large numbers of the interneuronal connections called *dendrites* (from the Greek *dendros*: tree). Each dendrite in turn has unusually large numbers of structures called *dentritic spines* that connect to other nerve cell projections called *axons* on nearby neurons. Neurobiologists, in a nod to horticulture, call this dendritic sprouting *arborization*.

Neuronal arborization is especially elaborate in the human neocortex, the outermost layer of the cortical gray matter that covers much of the surface of our brains. And it reaches its extreme in the frontal cortex, which influences personality and decision-making.

As our brains grow in size and mature, our neurons arborize furiously instead of multiplying. But other important brain cells called *glia* actually do increase in numbers. These glial cells surround the neurons, feeding them and acting as electrical insulators. They reach especially high numbers in our frontal cortex [11].

These huge increases in neocortical arborization and in the numbers of glial cells mark us out from all our primate relatives. Even our closest living relatives, the chimpanzees and bonobos, do not show such increases [12].

Remarkably, these uniquely human changes are not costly in terms of energy consumption. It is true that neurons in the human neocortex use up more energy than those of our primate relatives, because of their massively increased numbers of connections. But the fact that the neurons are more thinly distributed means that our brains actually use less energy per unit of volume than those of chimpanzees [13]. Our huge brains are costly, but not prohibitively so.

Human brains exhibit extreme versions of these developmental changes, but they build on developmental pathways that were already present in our mammalian relatives. It has been known for decades that when a rat is raised in an enriched environment the mass of its cerebral cortex increases. This environmental stimulation has a remarkable resemblance to the changes that take place during the development of human brains. Just as in the human brain, the increase in the mass of the cortex of stimulated rats has been traced to an increase in the number of glial cells, rather than to an increase in the number of neurons [14]. And recent studies have shown that an enriched environment also stimulates arborization in many of these rat neurons (Figure 81) [15].

A rat lucky enough to be raised in an enriched environment can increase its brain weight by as much as ten percent [16]. Similar increases are seen in parts of the brains of fruit flies that have been raised in what is (for a fruit fly) a

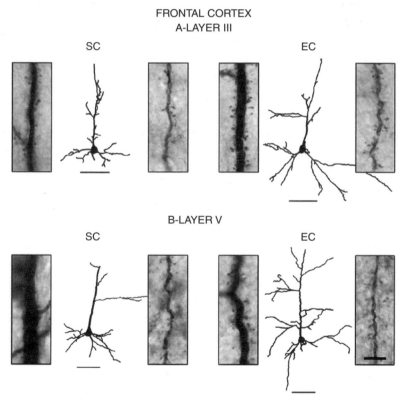

Figure 81 Typical neurons from the prefrontal cortex of rats raised in deprived (left) and enriched (right) environments show how arborization (dendritic branching and numbers of dendritic spines) increases when the environment is stimulating. The glial cells that surround the neurons are not visible in this picture, but glial cells also respond to a stimulating environment by increasing in numbers. From Figure 2 of [15].

stimulating environment. Flies that are raised in isolation in culture tubes have smaller "mushroom bodies" in their brains than those that are reared in groups and that can interact with each other [17]. The mushroom bodies of insect brains, like the outer cortical layers of mammalian brains, are involved in learning.

Similar experiments are of course impossible to carry out in humans. But brain imaging techniques may soon give us the ability to quantify the amounts of arborization and the numbers of glial cells, among many other parameters, in living human subjects [18, 19]. These techniques may eventually allow us to measure the effects of environmental changes, such as different teaching methods, on the very structures of our brains.

The mammalian brain is able to produce these cellular changes because it has a high degree of *neuronal plasticity*. This ability, which allows the brain to change as the environment changes, has deep evolutionary roots. Illustrating the ancient origins of this plasticity, developmentally plastic neurons have recently been discovered in fruit flies [20].

The increased arborization of human cortical neurons, and the large increases in the numbers of glial cells, are most apparent in the very parts of our brains that have undergone the greatest overall size expansion during our recent evolution.[1] What we don't yet know is whether these unusual features of human brains are enhanced even more in those of us who are raised in stimulating environments, in the way that they are in rats. If our brains do respond in these specific ways to stimulation, this would be a demonstration that developmental pathways leading to neuronal plasticity have become an important part of the genetic instructions for the human brain.

The sources of these structural changes in the brain are beginning to be discovered. As I was writing this chapter, my colleague Franck Polleux from the nearby Scripps Research Institute told me about some astonishing work that has emerged from a collaboration between his group and a group headed by Evan Eichler of the University of Washington. The two labs have now discovered some of the molecular and evolutionary mechanisms that underlie these neuronal changes.

Polleux was intrigued by the molecular biology and evolution of a gene called *SRGAP2*. He and others had found that this gene is turned on full-blast in the brain, where it contributes to the maturation of neuronal dendritic spines. Working independently, Eichler was exploring the role of gene duplication in human evolution. An unusual number of pieces of chromosome have become duplicated in the human lineage, adding about four percent to the size of our genomes compared with those of chimpanzees. Eichler wondered whether these human-specific duplications play a role in the development and adult function of human brains.

With the discovery that the *SRGAP2* gene has been caught up in these duplications, the two labs began to collaborate. They found that about three and a half million years ago a duplication of part of chromosome 1 caused a truncated extra copy of the *SRGAP2* gene to appear in the genomes of the hominan lineage that would eventually lead to humans. Then, a million years later, this gene copy was duplicated further to give rise to two more copies. The gene pools of modern humans, Neanderthals, and Denisovans all carry the three new copies along with the original gene, while our great ape relatives have only the original gene [22].

Only one of these extra copies currently makes a protein, but it is so damaged that it has lost the function of the original protein. When this defective protein is synthesized in the cell, however, it is able to bind to the functional protein that is made by the original gene. The binding interferes with the functional protein's activity. And the interference has surprising effects.

Polleux examined the effect on development of a nonfunctional SRGAP2 protein. He and his co-workers started by inactivating the *SRGAP2* gene in mice. As the mouse embryos grew there was a profound slowing of spine maturation, which led to a significant increase in the total number of spines that were formed by cortical neurons. When he inserted the human-specific defective form of *SRGAP2* into mouse cortical neurons that were growing in tissue culture, this also resulted in a massive increase in the total number of neuronal spines. The interference by the human defective protein had the same effect as inactivation of the mouse *SRGAP2* gene [23].

In the presence of the inhibitory protein, the development of the mouse neuronal dendrites was slowed, so that they remained immature for longer. This gave them time to make more spines and undergo more arborization. It is just such differences that distinguish the development of the human brain from that of the chimpanzee (or the mouse), in which brain development is not influenced by a defective *SRGAP2* protein.

Does this defective *SRGAP2* gene have the same effect in human neurons as it does in mouse neurons? If it does, then Polleux and his coworkers will have begun to unravel an important developmental difference that distinguishes our brains from those of other mammals. The defective *SRGAP2* protein would be exerting a crude kind of developmental regulation, slowing the development of our brains and allowing them to become more complicated in the process.

These experiments have only addressed what happens when a few neurons express this human-specific form of *SRGAP*. We await with bated breath the results of the next logical experiment. What happens when copies of the damaged *SRGAP2* genes are expressed in mouse embryos, and the mice are then grown to maturity? What effect will the suppressor proteins have on the development of their brains? Polleux suggested to me that the result might simply be disruptive, like putting a Ferrari engine in a Ford Fiesta. At the other extreme, the experiment might produce a strain of mouse Einsteins.

Copies of all three of these duplicated genes have spread through the human gene pool, but only the one that interferes with the original gene's function has become fixed so that everyone carries it. This raises the possibility that the successive duplications may have had a cumulative effect on brain function. Collectively, they may have contributed to a gradual lengthening of the critical period of brain maturation in the course of the evolution of the human lineage, helping our brains continue to mature beyond infancy into childhood.

Such cumulative effects on brain function may explain why each of these three gene duplications increased in numbers in the populations of our ancestors. As I write this there is as yet no direct evidence at the DNA sequence level for such selective sweeps. But, because Neanderthals and Denisovans also carried these duplications, detailed comparisons of their genomes with ours will soon reveal more details of the evolutionary history of the *SRGAP2* gene.

This regulatory change is a remarkably clumsy one, probably the result of the rapid evolution of human brains. Given enough time, more sophisticated regulatory mechanisms would probably replace this awkward way to regulate the *SRGAP2* gene. But the effect of this and other genetic changes is that our ancestors acquired new brain functions during a time-span that in evolutionary terms was almost instantaneous. They immediately began to make use of them.

We can now begin to understand one way in which our recent evolution has caused many of our cortical neurons to radiate out into those unusually dense dendritic arborizations. I expect we will also soon track down the genes that are responsible for the increases in the numbers of glial cells that provide the nourishment for all this neuronal development [24].

These cellular processes happen to all of us as we mature, and almost certainly they can receive a boost if our environment is stimulating. In effect, however clumsily, evolution has provided us with a huge built-in ability to increase our intellectual potential. It is inexcusable for us not to use it.

The Nuts and Bolts of Brain Evolution

Ultimately, evolutionary changes in our DNA have brought about the increased complexity of our brains. What were those changes? Did they take place mostly through changes such as those extra copies of *SRGAP2*, or did some of them happen in other ways?

As we have seen, genes are stretches of our DNA that carry the codes for proteins and for important RNA molecules. One of the most surprising pieces of information to emerge from the Human Genome Project is that we have embarrassingly few protein-coding pieces of DNA. Argument continues over the exact number, but it seems to lie somewhere close to 25,000.

To add injury to insult, it turns out that many of the genes that we do have are defective. A recent survey of the complete genomes from 185 humans found that each person in the sample carried defective alleles at an average of 100 of their genes. About twenty alleles in any given individual are so damaged that they cannot function at all [25].

Luckily, defective alleles are often recessive, so that they are masked by functional alleles that we inherited from our other parent. But some of us inherit recessive defective alleles of a gene from both of our parents, and these unlucky individuals may die or fail to reproduce.

This frighteningly high burden of defective mutant alleles may help to explain why we have so few genes. If we had twice as many genes, we would have twice the chance of being homozygous somewhere in our genomes for damaging recessive alleles.

It is probably not a coincidence that, like humans, many other large multicellular animals and plants have somewhere around 20,000 to 30,000 genes. This number may represent an upper limit, beyond which accumulating mutations are likely to have a real impact on survival.

This burden of harmful alleles may explain why we have so few genes, but it does not explain how we have managed to use these seemingly slender genetic resources to build our complex and multi-talented bodies and brains. Although all of us started as a tiny fertilized egg, we have been able to develop into an adult with 10^{13} cells and 200 different types of tissue. And all of us have brains of unparalleled complexity, with almost a hundred billion neurons that can communicate with each other through almost a quadrillion (10^{15}) interneuronal connections.

The answer to this much deeper puzzle about how we can do so much with so little is that most of our genes have evolved to become extremely versatile.

Because they are able to play many roles on the stage of life, their effective number is multiplied to far more than 25,000.

Our genes are so versatile because they have coevolved with other parts of the genome that regulate how they function. These additional genomic regions add new dimensions to each gene, in the same way that different software programs add functionality to the basic operations of a computer. Our genes make up only about one percent of our genomic DNA. The rest of the DNA codes for other stuff. Regulatory DNA sequences bind to regulatory proteins that turn various genes on and off, and other DNA regions govern how our chromosomes fold and unfold. Our cells also possess huge stretches of apparently pointless DNA that may simply be along for the ride (or that may be doing subtle things that we have yet to discover).

Each of our genes, with its equipage of regulatory DNA regions, plays a role in a developmental network that includes many other regulated genes. And each gene is subject to different patterns of regulation in different parts of our bodies. The genes in different tissues and organs are translated into protein in different ways and turned on or off to different degrees. As our bodies grow and develop, genetic regulation produces a different mosaic of gene activity within each of our cells.

The dystrophin gene provides a good illustration of how regulatory mechanisms can add to genes' versatility. This gene is located on the X chromosome (see Chapter Six). It is expressed in muscle cells, and also in the brain in neurons and glial cells. Mutations in different parts of the gene can give rise to the tragic muscle-wasting disease Duchenne muscular dystrophy, to mental retardation, and in the severest cases to both. Because the X chromosome is present in only one copy in males, the majority of the dystrophin disease burden falls on male children.

When dystrophin protein molecules are fully functional, as they are in most of us, they play different roles in different tissues. In muscle cells the dystrophin protein forms an essential part of the structure of muscle fibers. Dystrophin also helps glial cells to stick together, and it forms part of the synaptic contacts between adjacent neurons [26]. This protein's versatility is the result of regulatory mechanisms that determine which parts of its gene are transcribed into RNA and then translated, with the result that there are different versions of the dystrophin molecule in different cells.

If a region of the dystrophin gene that is only expressed in muscle cells carries a harmful mutational change, the result is muscular dystrophy. If mutational changes happen in other parts of the gene that are expressed in the brain, the result can be either brain dysfunction or a combination of brain dysfunction and muscular dystrophy.

Molecular biologists are gradually sleuthing their way through such regulatory complexities. In the process, they are finding out how gene regulation actually works. They have discovered that many regions of regulatory DNA provide binding sites for a whole zoo of *regulatory proteins*. Sometimes the regulatory proteins crowd onto every part of the DNA sequences that lie near the gene. They can even pile on top of one another. These proteins are coded by a diverse collection of *regulatory genes*.[2]

Each of the regulatory genes is itself subject to equally complex webs of regulation. These webs are often triggered by factors from outside the cell that cause a cascade of regulatory events (Figure 82) [27].

Because the mechanisms of gene regulation are so complex, they provide many ways by which mutations can change the pattern and timing of gene expression. And they also provide a huge target for evolutionary change.

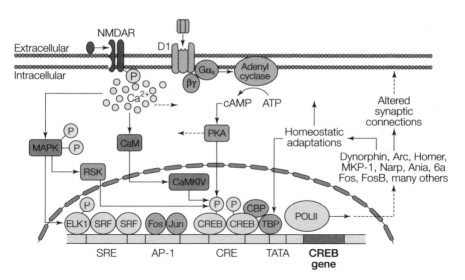

Figure 82   Even though we have relatively few genes, many of them are able to do different things in different tissues. This picture, which is modified from Figure 1 of [28], gives you a birds-eye view of some of the complexity of gene regulation. The bottom part of the diagram shows how regulatory proteins, including proteins made by the gene itself, are clustered on different parts of the regulatory DNA sequence that lies upstream of the *CREB* gene (lower right). (*CREB* is a gene that, among many other things, influences the modifications that take place in neurons during learning.) The top part of the diagram shows how the synthesis of these regulatory proteins is influenced by signals from outside the cell. Cells in other tissues of the body are bathed in a different external environment, which causes them to regulate their *CREB* genes in a different way.

About a quarter of our genes are expressed in the brain. That is, their information is transcribed into messenger RNA molecules and translated into protein in brain cells.

Thus, at first blush, it might seem reasonable to suspect that the rapid evolution of our brains is the result of selection-driven changes in these brain-expressed genes. After all, it is these genes that code for brain cell proteins. You might then be tempted to conclude that the genes that are expressed in the brain should have evolved more quickly in the human lineage than in the lineages that led to the chimpanzees and our other relatives. But, surprisingly, this seems not to be the case. Even *SRGAP2* has not changed much—though as we just saw its regulation has. Most brain-expressed genes do not evolve unusually quickly in humans compared to other primates [29].

What is going on? On the one hand, there is nothing remarkable about the rate of evolution of the genes that are expressed in the human brain. On the other hand, it is abundantly clear that our brains have been evolving like crazy.

Perhaps the speedup in our brain evolution can be traced, not to changes in the genes themselves, but to changes in that much larger evolutionary target, the huge stretches of DNA that regulate them.

It is difficult to test this hypothesis directly, because the regulation of each gene poses a set of unique problems and may take years of work to understand. Compounding the difficulty, there are at least forty million differences between the genomes of humans and chimpanzees. These changes range from alterations at single bases to big insertions and deletions. Most of these differences are likely to have little or no effect. This makes it hard to determine which, if any, of the changes that have taken place in known or suspected regulatory regions have actually influenced the evolution of our brains.

We can, however, employ an indirect measure of the rate at which our regulatory changes have evolved. It is now possible to quantify the amounts of the many different messenger RNA molecules that are made by a particular type of cell. This proxy measure can be used to provide a rough idea of how quickly gene regulation has changed during our evolution.

The measure depends on the fact that messenger RNA molecules serve as intermediates between the genes and the proteins that they code for. You will recall that these DNA-like molecules are transcribed from the genes inside the cell's nucleus. They carry the information that is coded in the genes into the cytoplasm of the cell, where the proteins are made.

Except for a few specialized types of cell that have lost their nuclei, most of our cells each carry two complete sets of chromosomes, one from each of our parents. This means that each cell possesses two copies of every one of our 25,000 genes (males have only one copy of the genes on the X chromosome, and a few addition genes from the Y). Thus, in theory, all 25,000 of our genes should be able to produce messenger RNA molecules in every cell. But suppose we find that there are no RNA copies of a particular gene in a given type of cell. This means that the gene, even though it is present in the cell, has not been turned on. Alternatively, if we find many RNA copies of a gene, this is a sign that the gene has been turned on full-blast.

A group of German and Chinese scientists [30] has compared the rate of RNA synthesis in thousands of genes in the brain cells of humans with the rates in those of our primate relatives such as the chimpanzee and the rhesus macaque monkey. The technique is so sensitive that it only needs a tiny sample of tissue from each species.

Measurement of all these different RNA levels depends on high-tech molecular methodology, in this case silicon microchips to which bits of DNA have been bound. Microchips can now be manufactured that carry bits of DNA from thousands of human genes. Each of these pieces of gene is bound to a specific location on the chip. When these chips are bathed in RNA from a particular type of cell, the amount of RNA that binds to each location on the chip can be measured. RNA from different types of cell will yield different binding patterns on the chip, because the mix of messenger RNAs in each type of cell is different.

This methodology enabled these groups to ask whether, relative to macaques, the pattern of expression of brain-expressed genes has changed more rapidly in humans than in our close relatives, the chimpanzees. If it has, then this provides indirect but strong evidence that there has been a speedup of regulatory changes in our lineage.[3]

They found that, using the amount of divergence from the macaque lineage as a control, the regulation of human genes has indeed changed more rapidly than that of chimpanzee genes. The human speedup has been approximately four-fold in the prefrontal cortex of the brain, and a little less in the cerebellar cortex. It is most marked in genes that show not only regulatory differences between the species but also differences from youth to old age. There has been no similar speedup in so-called "housekeeping" genes that are involved in basic cellular functions.

The authors also used additional techniques to make direct measurements of the rates of change in some regulatory regions, and in the genes that code for the

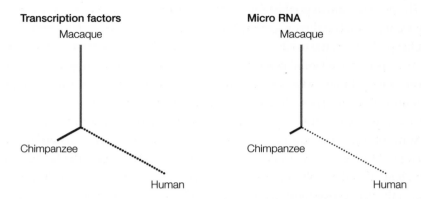

Figure 83 The evolution of regulatory regions and small regulatory RNA molecules in the human lineage has sped up since the common ancestor of humans, chimpanzees, and macaques (marked with an X in the figure). (Modified from Figure 3 of [30].)

tiny RNA molecules that also play an important role in regulation. Both of these sets of data also showed a speedup of the rate of evolution in the human lineage (Figure 83).

In retrospect, we should not be surprised by this finding. The size of the "target" for mutational changes in regulatory regions is much larger than the target that is presented by the genes themselves. Regulatory changes can arise not only in the thousands of DNA bases that flank the gene itself, but also in many other parts of the genome. And it may be that such mutations in regulatory regions are more likely to survive than mutations in the genes. It is possible that the web of regulatory interactions is so redundant that changes in regulation are less likely to be really damaging on average than mutations that affect the genes themselves. This conjecture has yet to be tested, however.

As a Species, Our Summed Brainpower is Evolving into a Rainforest of the Mind

So far, in this brief exploration of how our brains have evolved, I have emphasized how complex regulatory mechanisms have contributed to the development of all our brains. But now I want to introduce another layer of complexity, one that takes into account the polymorphisms in our gene pool that are maintained by genetic equilibria.

The present human population has seven billion pretty good brains. You may not want to dwell on this image, but if our brains were all piled up into great slithery heaps their volume would be equal to three and a half Great Pyramids of Giza.

As a species, we would labor under an enormous disadvantage if all of our seven billion brains were exactly the same. Luckily they are astonishingly, vibrantly different. These differences can be traced in part to the different combinations of allelic forms of genes and regulatory regions that we all carry.

Some of these genetic differences among us are likely to be maintained in our gene pool by frequency-dependent selective forces. This raises the fascinating possibility that the evolution of our intellectual talents may resemble the evolution of the very ecosystems that we must use those talents to save!

Of course, many of the most dramatic alterations in our brains have not been frequency-dependent. They have resulted from that mainstay of evolutionary change, directional natural selection.

Directional natural selection can drive highly favorable regulatory mutations that affect brain function to higher and higher numbers in our gene pool, where they eventually replace older alleles. It is apparently not a coincidence that a gene called *GPR56*, which is involved in brain development, acquired an extra piece of DNA at the same time as the cerebral cortex first appeared in mammals [31]. And, as we have just seen, duplicated copies of the *SRGAP2* gene have swept through the ancestors of Denisovans, Neanderthals, and modern humans, causing a reduction in the activity of the original *SRGAP2* gene.

These alleles have swept through the entire human gene pool, which means that they have affected all our brains. But there are many genetic differences between individuals that are not "fixed" in our population. Some of these differences contribute to differences in brain function among the different members of our species.

A thoroughly accepted way to measure these differences is to obtain intelligence quotient (IQ) scores and standardize them according to age. The degree of resemblance between IQ scores of related people can then be used to measure the contribution of genes to this character. This measure is IQ's *heritability*, the degree to which this characteristic is inherited from one generation to the next. If the heritability of IQ is high, genes play a large role in its expression; if it is low, environmental factors predominate.

Heritability is a statistical estimate of the contribution of genes to physical and behavioral characters. It can be obtained in a variety of ways, but it must always be interpreted with caution.

For example, we might search out a large number of pairs of identical twins and measure some character that they share, like height or IQ. This collection of data can then be used to derive a *correlation coefficient*, which measures the tendency for the members of each pair of twins to resemble each other. A high value for the correlation coefficient would indicate a high heritability. Similar calculations can be carried out on pairs of siblings, or on sets of comparisons between offspring and the average value of their parents.

But these correlations are also influenced by genetic and environmental factors that vary from one type of measure to another. For example, measurements of the similarities between identical twins yield artificially high estimates of heritability. This is because the twins share combinations of genes that help to make them similar, but these combinations break up in unpredictable ways when the genes are passed on to the next generation.

Heritability estimates from non-identical twins also tend to be substantially higher than those between siblings, even though non-identical twins and siblings have the same genetic relationship. In part this is because the twins experience the same environment in their mother's womb, while the pairs of siblings are nurtured by the same mother at different times.

Parent–offspring measures, too, can be influenced by the environment if the environment changes from one generation to the next. And heritability can lose much of its meaning if the environments are really different. Imagine a comparison between Tarzan, who was raised by apes, and his hypothetical twin brother who was raised back in England. Their resemblance would undoubtedly be lower than if Tarzan's biological parents had not traveled to Africa and had instead raised Tarzan in England—but how much lower, and why?

These difficulties have turned the study of the heritability of IQ into a minefield. The earliest of these studies used identical twins and yielded heritability estimates of about eighty percent. Such a high number implies that much of the tendency to score well or badly on an IQ test is coded in our genes and will be inherited by our offspring. But these early estimates were clouded by statistical problems and in some cases by outright fabrications of data [32]. The possibility that there is a genetic source for racial differences in IQ scores has sparked especially ugly debates [33].

Without revisiting all this ugliness in detail, let me cut to the chase. Current estimates of the heritability of IQ tend to fall at around fifty percent. If the effects of the prenatal environment are included, heritability can actually fall to below fifty percent [34]. Such a level of heritability means that genes are indeed important, but that environment plays at least an equal role. And we are also beginning to learn that genetic and environmental effects are not independent but can interact in unpredictable ways.

For example, consider the astonishing Flynn effect. In 1987, psychologist James Flynn reported that IQs have increased dramatically in the developed world over a span of mere decades [35]. Similar increases have since been found in the populations of underdeveloped countries [36]. These changes have taken place so rapidly that they cannot be explained by evolution. They must result from changes in the environment, which have influenced the pretty good brains that we have inherited from our ancestors.

During the century since the invention of IQ tests it has been necessary to re-standardize the tests repeatedly, in order to compensate for these increasing average scores. Richard Nisbett of the University of Michigan and his colleagues have pointed out if people in the United States in 1917 had been given today's tests their scores would have been as low as the lowest scores in the developing world today [37]. But there has been no change in potential brain-power since that time. If people who took the test in 1917 had instead grown up in today's world, their performance on IQ tests would be at today's levels.

The Flynn effect seems to be traceable to increases in environmental complexity, improved nutrition, and probably the spread of childhood immunizations. Even though the Flynn effect's gains seem to be leveling off in some European countries, in most of the world the function of people's brains has only begun to benefit from such improvements in the environment.

In particular, we are only just beginning to appreciate the large role that disease plays in depressing intellectual ability. Throughout the developing world, there is a highly significant negative correlation between IQ and the numbers of parasites that people carry in their bloodstreams and livers [38]. Tragically, and preventably, high burdens of disease have lowered the IQs of hundreds of millions of people worldwide.

Evidence continues to grow that environment and genes work synergistically in their effects on brain function. For example, eighty-five percent of European children carry either one or two copies of a particular allele of a gene that modifies dietary fatty acids. Within this group, some of the children were breast-fed as infants. They scored six to eight IQ points higher than those that had not been breast-fed [39]. This is a huge difference, amounting to half a standard deviation of the IQ distribution of a typical population.

Some of this difference may be explained by other factors like socioeconomic status. But it is striking that the other fifteen percent of European children, who do not carry this allele, do not show a raised IQ from breast-feeding, even though socioeconomic differences and other factors must be affecting them similarly.

This remarkable study illustrates the complexity of gene–environment interactions. It also raises two obvious questions. First, can the benefits of breast-feeding be extended to that fifteen percent of children as well, through some yet-to-be-discovered manipulation of the environment? Second, why do not all people carry this allele? Has it not had time to sweep through the population, or are there complicated selective forces that keep it in a balance? If so, then there may be a down-side to this allele that has yet to be discovered.

There are times when improvements in the environment are powerless to undo genetic damage. Mutant alleles have been discovered in almost three hundred different genes that can cause severe harm to the brain [40]. Each of these genes is therefore essential to brain function and development.

In a handful of cases, such genetically caused severe brain damage can be reversed. The harmful effects of the genetic disease phenylketonuria can be largely ameliorated if the amino acid phenylalanine is removed from the diet. But damage to most of these three hundred genes cannot yet be reversed by medical treatment or by altering environmental factors—though therapies involving the replacement of defective genes or cells are on the horizon for some of them [41].

Such badly damaged mutant alleles form one extreme end of the spectrum of genetic variation that contributes to brain function. The rest of the variation, most of which tends to have much smaller effects, is also much more likely to be influenced by the environment. This strongly interactive genetic variation influences not only IQ, which has received most of the press, but also the many other components of intelligence that provide our species with such a wide variety of talents [42].

Recent advances in gene sequencing have given us new tools to search for such variation. Massive DNA sequencing projects, including the intensive sequencing of the genomes of a thousand different humans, have uncovered millions of genetic markers called SNPs (single nucleotide polymorphisms, pronounced "snips"). These are specific sites along the DNA double helix at which more than one base is found in the population.[4]

SNPs are scattered plentifully through all our chromosomes. Some of the alternate bases at SNPs are common and widespread, suggesting that these polymorphisms have been present in our gene pool during all the time that we evolved in Africa and spread throughout the rest of the world. But the majority of the alternate bases found at SNPs are rare, and the majority of these rare bases tend not to be shared among different human groups [43]. Most of them are likely to be recent arrivals on the genetic scene.

How many of these millions of SNPs actually influence brain function? Probably only a relative handful. But SNPs can be used as genetic markers to follow the fate of the genomic regions that are tightly linked to them. And alleles that are present in some of these linked regions are likely to have an effect on brain function and development.

There have been numerous attempts to search for associations between the presence of particular alleles at SNP sites and brain function, but so far these surveys have found only weak signals that tend not to be repeatable from one study to the next [40, 44, 45]. This may be because most of the alleles that influence brain function tend to be strongly influenced by the environment, weakening any correlations between the genes and their effects. Another limitation is that up to this point most scans of the genome have been confined to the relatively small proportion of SNPs that are shared by many of the people in the sample. Genes that are linked to rarer SNPs may be more informative, but they are much harder to study because they require huge sample sizes from within particular human groups.

The hunt for the genes involved in IQ continues. In the meantime, another way to investigate the evolution of our brains is to ask whether there are signs that pieces of DNA that affect brain development have recently been dragged up in numbers by natural selection in the human population. Such strongly selected alleles should carry with them clusters of closely linked genetic markers. Because of genetic recombination, the signs of such sweeps soon fade away, so that we cannot detect them if they are more than a few tens of thousands of years old. A number of interesting genes have emerged from these studies, but few of them can be directly connected to the function of the brain [46, 47].

Because Neanderthal genomes are now available for comparisons with ours, a handful of alleles that seem to have been driven up in our species by selection since human–Neanderthal divergence have also been found [48]. Again, however, few of them have obvious associations with brain function.

Even if the genes that influence brain function are frustratingly hard to find, it is clear from the substantial heritability of IQ that they must be everywhere in our genomes. Each of us carries a different combination of these thousands of alleles. This variation, along with a huge input from our ever-changing environment, explains why our billions of brains are so different from each other.

Eventually, many alleles that influence how our brains function will be discovered. But, as we have seen, genes are only part of the story. As our environment changes and becomes more complex, individuals of each generation will be able to realize more and more of their intellectual potential. And the genetic variation in our species as a whole is being reshuffled and expressed each generation in new and unexpected ways. This is especially the case in recent decades, as divergent groups of people bring their differing genetic legacies to the great mixing bowl of genes that is becoming characteristic of modern humans.

I suggested more than a decade ago that some of this genetic variation might be maintained by negative frequency-dependent selection [4]. As alleles that add new capabilities to our brain function accumulate in a population, some of them may have an advantage when they are rare but lose that advantage when they are common. Over time, just as happens to species in ecosystems, these alleles will reach equilibria in our gene pool. This is how we are evolving our teeming rainforest of the mind.

In the search for the sources of our rich and complex intellectual heritage, it is essential to be alert to the possibility of such frequency-dependent selection. And, because brain function is so complex, it is likely that different human groups have taken somewhat different paths in the accumulation of the genetic variation that contributes to their intellectual talents. But it is well to remember the amazing inventiveness of the peoples of New Guinea, of the first migrants to the Americas, and of other groups throughout the world. Such inventiveness clearly shows that there are many equally effective paths to the acquisition of pretty good brains.

How are we using all of these marvelous resources of ours? In this book we have traced the stories of the first peoples who ventured into the high valleys of the Himalayas, those who traversed the vast rainforests of Southeast Asia and arrived in the remote lands of New Guinea and Australia, and the bands of hunters who spread across the boundless prairies and forests of the Americas. Theirs are stories of resourcefulness and survival, but none of these migrants had any notion of the true potential of the kilogram-plus of gelatinous tissue that they carried in their heads.

Can we do a better job than these pioneering groups of people of coming into balance with our environment? If we fail, we will have wasted the collective potential of our seven billion pretty good brains. It is this central question that I will explore in the last chapter.

CHAPTER TWELVE

Green Equilibrium is More Than a Metaphor

W e are now nearing the end of our search for the processes that govern the world's ecosystems and our role in them. During our journeys we have traveled around the planet and plunged into the depths of the oceans. We have even braved the depths of the human gene pool.

We have followed the flow of energy from the sun through plants to animals, and discovered what happens to ecosystem diversity when that flow is altered—for example, by our indulgent support of a vast population of predatory domestic cats. We have encountered the many frequency-dependent evolutionary and ecological forces, ranging from predator–prey interactions to host–parasite relationships, that have contributed to ecosystem diversity. And we have learned that all the members of an ecosystem, including the organisms that make up that invisible world of pathogens, parasites, and tiny symbionts, play a role in maintaining stability and diversity. The tiny beetles that are essential to the reproduction of the giant water lilies of tropical American rainforests also contribute to the survival of the vast collection of plants, birds, and fish that interact with the water lilies.

Natural selection is a process of creative destruction. It is of course a tragic irony that millions of animals and plants must die every generation in order to maintain the health of each ecosystem. These deaths cannot be random, or the ecosystems will break down. For a green equilibrium to emerge, many of the deaths must result from natural selection. And much of this selection in turn will be the result of complex, frequency-dependent webs of ecological interactions among species, such as those that are beginning to be glimpsed in the New Guinea forests by Vojtech Novotny and his colleagues.

These uncounted deaths, of creatures as diverse as bacteria and whales, have also contributed to the maintenance of genetic variation within each species. And that variation can in turn permit a few members of some species to survive

Figure 84 Filip Damen, who brought together eleven clans to protect Papua New Guinea's lowland forest, greets the dawn in his village. Behind him stretches some of the forest that his efforts have preserved from logging.

even existential threats. Recall the handful of ancestors of the arapaima that managed to survive by burying themselves in the mud as their world was burned and blasted into oblivion.

The green equilibria that underlie these evolutionary and ecological processes are not metaphors. The numerous frequency-dependent balancing acts that make up these equilibria maintain both ecological and genetic diversity. They also make it possible for future ecological and evolutionary change to take place. This is why I pointed out, at the start of our adventures together in this book, that we destroy these equilibria at our peril. It is not just ecosystem diversity that will be lost, but also the possibility of change and survival in the future.

Until recently, our own species was fully caught up in these green equilibria. We suffered high levels of mortality along with every other species. Our numbers were regulated by the predators that caught and ate us, and by innumerable frequency-dependent interactions with our invisible world of pathogens and parasites. Even our emerging technology, which was driven by the Darwinian brain–body–genes–environment feedback loop that we explored in Chapter Eleven, did not remove these influences. Even as we were inventing this technology, frequency-dependent mortality continued to affect our species. Population explosions, aided by advances in communication, hunting, and agriculture, were followed by population crashes. Wars between tribes increased the language and culture differences between them. In New Guinea, different tribes have utilized different components of their natural world to enhance the effectiveness of sexual selection and their ability to attack other tribes. The result has been an explosion of cultural diversity on the island. Similar stories can be found in many other parts of the world.

But now our technology has reached a new level. It has suddenly knit our species into a global entity. We are all able to share a growing pool of scientific, technical, and medical knowledge, along with an avalanche of information about the state of our world.

In theory, this combination of new knowledge and rapid information transfer should enable us to chart a detailed and feasible course towards a restoration of our planet's ecological balance. Are there any signs that this is happening? We are bombarded daily with stories of the effects of heedless environmental destruction, and the disastrous consequences of our exploding population. But we are also keeping track of this damage as never before. A growing number of scientists and activists are beginning to take advantage of this new knowledge. I leave you with a few glimpses of these changes, and with some reasons for hope.

An Ecosystem on the Brink

In late 2010 I began a two-day trek into one of New Guinea's northern lowland rainforests. These were the same diverse forests in which Vojtech Novotny's group had discovered almost seven thousand different food-web links among the plants and the insects that feed on them (Chapter Two).

My goal was to reach the Wanang Plot, one of the newest parts of the network of forest dynamics plots that has been set up by the Smithsonian's Center for Tropical Forest Science. This network now includes fifty pieces of primary forest scattered across the planet. The Wanang Plot occupies a ridge in a remote region of low heavily forested hills that lies between the northern coast and New Guinea's central spine of mountains. Like the other plots, it is scheduled to be censused repeatedly in order to follow the dynamics of rainforest change. And it has deliberately been designed to be difficult to reach.

It was necessary to establish the plot in a remote area because tropical lowland forest constitutes one of the world's most endangered ecosystems. Around the coastal city of Madang, logging companies based outside New Guinea are rapidly chopping down huge swathes of this unique ecosystem. It took us half a day to drive from Madang to the head of the trail that leads to the plot. On the way we encountered dozens of speeding trucks loaded with gigantic rosewood trees. At one point, as we were coming around a bend, one of the trucks came directly towards us. It had taken the curve at top speed, and had begun to skid sideways on the gravel onto the wrong side of the road. Our driver swerved, and we were able to pass the truck on the right with centimeters to spare.[1]

We were still quivering from this near miss when the road suddenly ended. Our way was blocked by a great gash in the earth, filled with boulders and with the mangled remains of a bridge, which had been destroyed by a recent flood. The flood's destructive effects had been magnified by runoff from the logged areas that surrounded the road.

The gash marked the start of a two-day hike to the plot. The trail to the plot runs for more than half its length through open country that has already been logged. Some of this land is now being farmed or is starting to revert to second growth (Figure 85). The hike, through sun-blasted terrain bathed in fierce humidity, was

Figure 85   On the way to the Wanang Plot, near Papua New Guinea's northern coast, we stopped in one of the few patches of shade in a vast region of scrubby second-growth forest.

far more difficult than my hike a week later to the Ubii Camp through the champagne-like air of the highlands. But there were compensations—at one point on the morning of the second day we could see Mount Wilhelm, the highest peak in independent Papua New Guinea.

It was not until the middle of the second day that we left the cleared land and plunged into the shade of primary forest. The last ten kilometers of our journey crossed a series of heavily-forested knife-edged ridges. Although wildlife was difficult to photograph in the forest, we did manage to sight two wallabies and a cassowary.

The forested region was still intact in large part because of the efforts of a federation of local villages. On the way to the station I met the man who is chiefly responsible for this preservation effort. We had broken our trek in the village of Wanang, a third of the way to the plot. I slept that night in a thatched schoolhouse, which was the first school to be built in the area. The next morning I met Filip Damen, the Chief Magistrate of the cluster of villages around Wanang (Figure 84).

In 2001 the national Papua New Guinea government announced an agreement with some of the tribal landowners that opened up to logging the entire area of the middle Ramu River, more than 100,000 hectares. Damen persuaded some of the local landowners to opt out of the agreement. They insisted that their land be included in a Conservation Deed that was designed to set aside 10,000 hectares as a reserve, even though revenue from the concession would be lost.

With help from Novotny's Binatang Research Center in Madang, Damen obtained money from the logging companies for local jobs, healthcare, and schooling. The concessions wrung from the logging companies mean that the children of the villages are attending school for the first time. The schoolhouse I had stayed in was one of the results. The school has two teachers, and its 160 children come from the villages and from workers' settlements in the surrounding logged areas that we had just hiked through.

Perhaps most remarkably, Damen told me that he had taught himself to read and write during the tortuous negotiations. He also told me about his trip in 2009 to San Francisco, where he had been given the Seacology Prize. It was the first time that this modest but determined leader of his people had left Papua New Guinea.

Filip Damen and his villagers have been able to join together and overcome the millennia of distrust among tribes that has been exploited by outside corporations. They acted just in time to save an essential part of Papua New Guinea's northern coastal ecosystem. The forest, unlike the vast logged region to its east, is rich in bird, animal, and insect life. Hornbills flew overhead each morning of my stay, sounding like fleets of helicopters.

At the end of our trek lay the Swire Research Station, a set of buildings perched on stilts near the Wanang Plot. The station is there because of the villagers' determination. Their efforts and sacrifice have helped New Guinea's new generation of parataxonomists trace the web of ecological interactions that supports their rainforest home.

Some of these budding scientists are themselves only a single generation removed from the Stone Age. In one leap these descendants of hunter-gatherers have become scientists, determined to understand the laws that have shaped their world.

The parataxonomists are part of a new generation of ecologists who have taught us that we must pay attention to all parts of the web of life. The many plant–insect interactions that they are studying provide evidence that even lowly parasites play an important role in ecological stability.

It is obvious that restoration of ecological stability requires the re-establishment of predator–prey interactions among large animals, along with grazing and browsing interactions among animals and plants. But it is less obvious that stability also requires re-establishment of the many interactions between hosts and their parasites and between groups of mutualistically interacting species. The parataxonomists of New Guinea have helped to illuminate this essential component of the world's green equilibria.

Clearly, we have the capacity to heal our world and make it an even more diverse and well-balanced home for our remarkable species. But the speed with which the world is changing adds urgency to our task. We have seen how economic exploitation, warfare, and unrelenting poaching and hunting are emptying the tropical forests of Southeast Asia of their top predators and other essential members of the forests' food chains. And our visit to the wetlands of South America's Pantanal revealed how these rich ecosystems are only protected from exploitation because they flood for part of the year. The *Hidrovia* project could soon drain away this protection.

Shangri-La Loses Its Innocence

The tribes of New Guinea are being catapulted from the Stone Age to the twenty-first century in a single generation. In the Himalayan kingdom of

Bhutan the leap into modernity has not been quite so vast and disorienting, but in the space of decades the entire country has emerged from a medieval time-warp into the modern world (Figure 86).

What evolutionary and intellectual resources can the Bhutanese (and the rest of us) use to adjust to such rapid changes? During our visits, Liz and I discovered that the diverse peoples of Bhutan bring both strengths and vulnerabilities to this new world that they are entering.

In late 2009, we stood on a suspension bridge above western Bhutan's Burning Lake, while our guide Sonam Choki told us a little about the tangled skein of history and legend that has shaped his country.

Despite its incendiary name, the lake is simply a deep jade-green pool in one of the country's beautiful rivers, the Tang Chhu. The pool has been preserved as a shrine to the legendary exploits of an actual historical figure, Buddhist saint Pema Lingpa. Five hundred years ago, he is supposed to have plunged into the pool's swirling waters in search of a sacred text. The treasure had been hidden in the pool's depths six centuries earlier by the Guru Rimpoche, the founder of

Figure 86 These children from the Bhutanese district of Bumthang are on their way to a school near their village of Chamkhar. Before the 1950s there was little education in Bhutan outside of monastery schools. In 2004 Bhutan's literacy rate was still only 53 percent, the fifteenth lowest in the world.

Bhutanese Buddhism. To light his way, Pema Lingpa carried a lamp that miraculously kept on burning as he swam into the green depths. The glow of the lamp heralded his successful return, as he surfaced clutching the text.

Hundreds of colorful prayer flags fluttered above the shrine. We helped a group of pilgrims from Sikkim to string up even more, then drove a few kilometers along a gravel road to the nearby farming village of Dubler.

The largest farmhouse in the village is also the oldest, four stories tall with massive foundations of local stone. Each of its numerous rooms is steeped in a dark antiquity.

Nobody knows how old the house is. The village's written records have all been destroyed over the centuries in various accidental fires. But Dorji Lharno, the family's matriarch, told us that her family had lived there for at least five generations. Further, she claimed that Pema Lingpa, the free-diving saint of the Burning Lake, had himself once lived in the house. And indeed, a stone with an impression of the saint's foot occupies a prominent position in the house's shrine room (Figure 87).

To most of us, Bhutan still seems as remote and mythical a place as James Hilton's fictional valley of Shangri La. Although little is known about its early history, it is clear that its first settlers in the region were diverse. The present-day peoples of eastern Bhutan speak Sharchopkh, a language with affinities to those found in southern China and Myanmar. The major language of the western Bhutanese, Dzongkha, resembles languages of southern Tibet. The Layap, Brokpa, and other tribes on the frontiers with Tibet and China preserve a variety of other languages.

During the height of the Tibetan empire, from the seventh to the ninth centuries, Bhutan acted as a refuge for rebellious Tibetan monks and political leaders who had fallen from favor. Then, when Tibet's empire began to collapse, more monks fled to Bhutan. They brought with them architectural and engineering skills that enabled them to build Bhutan's great temple-fortresses, the Dzongs.

Warfare among Bhutan's tiny kingdoms left the region weak and unable to resist Tibet's power. All this changed early in the seventeenth century with the arrival of another exile from Tibet, the Shabdrung Ngawang Namgyal, who linked the kingdoms together to form Drukyul, the Land of the Thunder Dragon. The newly united kingdom invaded the flat Indian plains to the south, touching off an intermittent war with the British that lasted for a century. Although the Bhutanese managed to retain their independence, they were driven back into the Himalayan foothills.

Figure 87 In the shrine room of a farmhouse in western Bhutan, monks use cymbals, horns, and a skin-covered drum as they pray for the health of the family's absent son.

At the end of the nineteenth century the ruler of the central Trongsa valley, Ugyen Wangchuck, bowed to history and formed an alliance with the British. In 1907 he was crowned the first king of Drukyul. During his rule, which lasted until 1926, he opened Bhutan's first secular schools and its first hospitals.

Bhutan's cautious opening to the rest of the world continued during four successive reigns of Ugyen's descendants. This succession of strong kings encouraged a high degree of social cohesion. The Bhutanese have been able to institute environmental policies that would have triggered endless argument elsewhere in the world.

Bhutan's forest cover has increased by five percent between 1990 and 2005, partly as a result of the annual "Social Forestry Day." On each anniversary of the fourth king's coronation, thousands of citizens gather to plant trees (Figure 88). During the same period, nearby Nepal lost almost a quarter of its forest.

But at the same time that Bhutan was hesitantly opening up, the outside world was arriving with a vengeance.

Nepali immigration into Bhutan started early in the twentieth century. Although the immigrants were first welcomed as they filled jobs in construction

Figure 88A   I took this picture of Thimphu's Trashi Chhoe Dzong and the forested hills behind it in the fall of 2009.

Figure 88B   Later I found this photograph from 1935, taken from the same vantage (though with a wider lens). The picture shows extensive deforestation and erosion of the hills behind the Dzong.

and agriculture, many of them have recently been driven out of the country because they refused to adapt to Bhutanese culture.

And now the Indians are coming, lured by the expansion of Bhutan's greatest resource and chief export, hydroelectric power. Between our 2009 and 2011 visits, we found that the capital of Thimphu had been transformed from a peaceful town into a hectic construction site. New building projects were dumping wastes into the Thimphu River, which had flowed clear and green during our first visit.

The countryside was being transformed as well. In May of 2011 I visited the Thimphu office of Tashi Dorji, the head of Bhutan's Power Bureau. I told him how, a few days earlier, we had followed the Mangde River south from the great citadel of Trongsa through a wide, heavily forested valley that led towards the Indian border. On the way we had encountered many places where Indian crews were widening the roads, destroying entire tree-covered slopes and opening up the slopes to further erosion. When I asked about this, Dorji admitted that his government has little control over how the Indian contractors carry out their work.

Dorji told me that until recently all the hydropower development in the country had been in the form of ecologically sound "run-of-river" projects. This type of project is a sensible choice for Bhutan's rivers, most of which are small and flow swiftly. The rivers are not dammed, but instead some of their flow is diverted through tunnels to turbines.

In Nepal, collapses of several dams have caused the evacuation of millions of people [1]. Partly as a consequence of these disasters, Nepal is unable to supply enough electricity for its own population, much less export any to power-hungry India.

Dorji told me that ten new hydroelectric projects have been started in Bhutan in collaboration with Indian companies. The aim is to generate 10,000 megawatts by 2020. While most of the projects will be run-of-river, three of them near the Indian border will be conventional dams and reservoirs with all their attendant risks. Any damage, he pointed out, would of course be downstream of the dams on the Indian side of the border.

He also predicted that by 2013 there would be an influx of 100,000 new Indian workers into Bhutan, which currently has a population of less than a million. If each of these workers were to bring only one dependent with them, this would be the equivalent of 60 million new people arriving in the U.S. over a two-year period.

As Bhutan has opened itself up to the world it has benefited from increased education and health care. These in turn have nurtured a growing concern over the environment. And Bhutan has given the world much in return. In 1972 the grandfather of the present king, Jigme Singye Wangchuck, memorably coined the phrase "gross national happiness." This concept encapsulates the idea that a strong sense of community and tradition, along with access to education, health care, and equal opportunity, are more important than the gross national product. The idea has caught fire—in April of 2012 the U.N. General Assembly unanimously added the gross national happiness (GNH) concept to the global development agenda [2].

As the authors of the United States Constitution were well aware, for most people happiness remains elusive. A 2010 government survey of Bhutan's provinces measured nine different criteria that are likely to be important in happiness, ranging from health to community vitality and cultural diversity. The survey calculated that, based on these criteria, fifty-nine percent of Bhutanese ought not to be happy [3]. Bhutan scores lower in such happiness measures than many developed countries, primarily because of poverty and a continuing shortage of educational opportunities [4].

Happy or not, the outside pressures for development threaten to overwhelm tiny Bhutan. On both of our trips we drove and trekked through many parts of the Black Mountains, mere foothills of the Himalayas that would be substantial mountain ranges anywhere else. The mountains are still covered with dense forests that preserve the country's ancient green equilibria. But Bhutan will need all of its human resources, including the participation of all of its growing diversity of peoples, to retain these equilibria.

Nepal—Dysfunctional Government, Local Initiatives

Bhutan is leaving its time capsule and entering a period of rapid and disorienting change. In Nepal, invaders and warring empires have devastated the countryside for millennia, and a long series of more recent governments have been dysfunctional for centuries. As a result, the disruption of ecological equilibria has been far more serious. But, in spite of this general anarchy, we found that local initiatives and assistance by outside groups give some reason for hope.

Most of the people of Nepal live in the semitropical lowlands and the Kathmandu Valley. And Nepal, unlike remote Bhutan, has for millennia been directly in the path of great population migrations and collisions between empires.

The Kathmandu Valley, the much-fought-over site of many of these events, is an unusual area of level terrain in this great tectonic collision zone. The valley began as a precipitous set of gorges, carved out of the Himalayan foothills by the Bagmati River and its tributaries. Then, a little over two million years ago, earthquake-driven landslides formed a natural dam. The dam held back what would become the largest of many such ephemeral lakes in the Himalayas, with an area of 650 square kilometers [5]. Rich black alluvial soil washed down from the mountains and built up behind the dam.

After lasting a million years, the dam failed catastrophically. The lake persisted, though reduced in depth, until it finally dried up completely about 15,000 years ago at the height of the last ice age. The valley floor that emerged from the waters of the lake was rich and verdant.

A scattering of artifacts in the valley date back a mere three thousand years, though human occupation must surely have been longer. The first recorded kingdom was established by Kirati people from the northeast in approximately 700 BCE. Wars raged over the valley's abundance and kingdom succeeded kingdom, at the same time as Buddhism gradually gained a foothold.

During these upheavals Nepal was an essential part of the tea-horse trade route between India, Tibet, and China. The beautifully decorated temples and houses of Bhaktapur, an ancient city state and trading center a few kilometers east of Kathmandu, were built during this time (Figure 89).

The monarchy is now gone, after a three-century-long series of coups and assassinations. Nepal is currently one of the world's most unstable parliamentary democracies. Throughout the country, ethnic groups are demanding autonomy from the crippled central government. Twenty-four political parties are represented in the parliament, but many more lurk in the wings. I was bemused to learn that Nepal rejoices in thirteen different Communist parties. I had not imagined that there could be so many imperceptibly fine gradations of communism.

At this writing Parliament is controlled by a Maoist communist party. Before they came to power the Maoists had been fighting the government for twenty years, using terrorist tactics. Both my physiologist colleague Buddha Basnyat and our guide Raju Chaudhari told me, with wry amusement, that the Maoists had been voted into power because it was hoped that if they were given responsibility they would not be so violent. Political violence has indeed decreased, but promises of a more decentralized form of government that has room for Nepal's many minorities have not materialized.

Figure 89 The Nyatapola temple in Bhaktapur's Durbar Square, the tallest in Nepal, was built by a Malla king in 1702. A mix of Hindu and Chinese traditions, it has survived earthquakes and wars ever since.

Clearly, Nepal's central government, unlike the stable monarchy of Bhutan, has been dysfunctional for a long time. But, away from the anarchy of Kathmandu, I was struck by the resilience of other parts of the country.

The headquarters of Sagarmatha National Park, the park that includes Mount Everest, is set on a windswept hill above the town of Namche Bazaar. On clear days you can see Everest from the headquarters' grounds. There I met Rana Jit Gurung, a forestry supervisor who has been planting trees in the park's valleys for the last thirty-one years. He is determned to overcome damage that has been caused by centuries of deforestation.

In stark contrast to Bhutan, official estimates are that only about twenty-five percent of the original great forests of Nepal's foothills remain. But in fact this number is an exaggeration. Much of the forest that has survived is badly degraded by erosion and by the harvesting of firewood. Many of the areas listed as forest are degenerating into eroded scrubland. The U.N.'s Food and Agricultural Organization estimates that in reality only about ten percent of the original forests are left. Even these areas are laced with innumerable roads, paths, and trails that trigger erosion on the steep slopes. On top of this, fires are common in the remaining forests, damaging some of the precious stands of rhododendrons. Plate 19 shows one of the many magnificent clusters of rhododendron trees that we hiked through on the way to the Everest base camp, a vivid example of what is in danger of being lost.

The erosion has many consequences, all of them bad. Every year Nepal's rivers carry about 240 million cubic meters of precious topsoil down to the Indian flatlands. This loss has led reporter Erik Eckholm to comment ironically that topsoil is Nepal's most valuable export [6]. And topsoil is an export from which Nepal receives no compensation.

The erosion has other devastating consequences for Nepal's economy. Most of the hydroelectric plants in Nepal, as in Bhutan, are the run-of-river type, in which river water is diverted directly to turn generators rather than stored behind dams into lakes. If the water is clean and free of debris and the river flows are predictable, as they are in neighboring Bhutan, run-of-river plants work well. Alas, the silt- and debris-filled waters of Nepal's rivers often damage the generators and shut them down.

The hydroelectricity that neighboring Bhutan sends to India accounts for about fifty percent of all its exports, and is slated to rise substantially in the near future. Nepal must use almost all of its electricity domestically, and that is still not enough. Each night Kathmandu is afflicted by rolling blackouts. Neighborhoods and shopping areas are plunged into a profound darkness, lit only here and there by power from tiny portable generators.

Meanwhile, Rana Jit Gurung and his small crew of workers, with the support of the Hillary Foundation, have been planting trees on slopes in Sagarmatha that had been denuded by logging, fires, and landslides. They plant thirteen different species of firs, pines, and rhododendrons, and they have been so successful that he told me he is literally running out of places to reforest.

As we trekked up towards the base camp, we met many forestry patrols. The forests in this most-visited part of the country, including the dazzling rhododendron groves, seemed to be in excellent shape.

Even though the peoples of Nepal have been ill-served by their governments throughout their history, they have managed to retain a vibrant and diverse culture. In this they resemble the peoples of India to their south, who have survived generations of cruel and rapacious Shahs and Maharajahs and the plunderings of the British East India Company. And Sagarmatha stands as a beacon of what can be done by local experts and local enthusiasm with even a small amount of outside help.

Will the precious environments of Bhutan and Nepal survive the impact of their exploding human populations and the collisions of politics, tradition, and greedy exploitation? To save their environments, both countries must be able to draw on all their intellectual resources.

Bhutan's society is embarking on a period of rapid change, as immigrants flood into the tiny country. It is only now discovering the consequences of severe environmental degradation. Nepal, much closer to the crossroads of southern and central Asian cultures, has been subject to cultural invasions for millennia. And Nepal's environment has been degraded, not just for decades, but for centuries.

This long history of destruction makes it all the more impressive that Rana Jit Gurung and his small band of foresters have been able to reverse the damage in Sagarmatha. The replanting of the park's diverse forests, helped by the constant patrols of the area by park rangers, has also fostered the recovery of populations of snow leopard, blue sheep, and Thar goats. In 1970 there were no signs of snow leopards in the park, but foot patrols that were carried out in 2004 found more than fifty traces of these most elusive of the big cats [7].

Local populations, tourists, and the environment have all benefited from this re-establishment of a green equilibrium at Sagarmatha. And the park has also had luck on its side.

As we saw in Chapter One, endangered species have been able to repopulate Southern California's Sedgwick Reserve because they have refuges in remote regions of the coastal mountains that surround the Reserve. These refuges have provided the resources needed to restore the complexity of the park's predator–prey interactions, and perhaps the balance of its invisible world as well.

Remote refuges have played the same role at Sagarmatha. But now these refuges themselves are in danger. Rara National Park, and other remoter parts of western Nepal, are far from the established tourist treks. They are being mercilessly logged and plundered by poachers. Rara's few rangers are doing their best to protect the snow leopards and black bears that are being shot and poisoned by local hunters [8].

Back in Kathmandu, on our last day in Nepal, we talked with Ghana S. Ghurung, the Conservation Program Director for the World Wildlife Fund. He gave us a final glimpse of the byzantine history and politics that have bedeviled Nepal's hesitant entry into conservation. For seven years, WWF fought for the establishment of a buffer zone to the south of Sagarmatha that would effectively extend conservation efforts beyond the park boundary. Delayed by the recent murder of the royal family and the Maoist rebellion, the zone is now established and thriving with 26,000 visitors a year.

Further to the south, in the Terai, rampant logging that had once devastated the entire region, particularly from the time of Indian independence down until 1970, has been slowed. Rhinos have grown in number from fewer than 70 at the time of the founding of Chitwan Park to more than 700 today. Fish-eating crocodiles, once locally extinct, have been reintroduced. And Ghurung estimates, on the basis of automated camera records, that there are still 150 tigers in the area. But poaching continues. Some poachers have been caught four times and still go on poaching.

Finally, Ghurung told me that in Sagarmatha itself snow leopard mothers with cubs have recently been spotted. Blue sheep, the leopards' commonest prey, remain extremely rare in the park. As a result the leopards are starting to switch their search image and prey on Thar goats. I expect, if the sheep start to increase in numbers, that the leopards' attention will switch back from goats to sheep. The frequency-dependent forces that can lead to green equilibria continue to operate on the roof of the world, in spite of the machinations of politics and all the potentially tragic consequences of ignorance.

The Joys of Restoring Ecosystems

In this book we have explored the impact of our species on green equilibria around the world. We have also seen that the selective pressures that have shaped our species have also given us the means to preserve them. On one side are the destructive forces of greed, which often prey on the grim need for people to survive. On the other side is our growing ability to use our true intellectual powers, bringing together the power of our billions of brains for the first time in the history of our species, to restore the balance.

There is another factor in this equation—the sheer joy of helping parts of the world to survive. That joy is real, even if it cannot easily be quantified.

We have found examples of such joy everywhere in our travels. It can be found in abundance at Karanambu Lodge, in Guyana's interior. The lands around the lodge spread over 325 square kilometers of grassland, swampy lakes, and dry forest, and include a beautiful stretch of the Rupununi River. It was a working cattle ranch until 1983, when its owner Diane McTurk converted it into an eco-resort and research station.

Karanambu has been in the McTurk family for generations. Diane's great-great-grandfather Sir Michael McTurk, a medical officer, mapped the course of the Rupununi for the first time early in the nineteenth century. He returned to England, but his grandson, who was also named Michael, was drawn to

Figure 90 I encountered this young giant anteater, *Myrmicophaga tridactyla*, early one morning as it strode purposefully across the Rupununi savanna. Sixty years ago Tiny McTurk galloped across this same grassy plain and lassoed a much larger anteater for collector Gerald Durrell.

Guyana—British Guiana as it was then—and settled there. He became the colonial administrator of the native peoples for the country's central region. His son, Edward "Tiny" McTurk, started the Karanambu ranch in order to raise cattle and to collect a kind of super-rubber that oozes from the locally abundant balata trees. Diane told me that balata rubber is still in demand for the most expensive golf balls.

The first moment of fame for the ranch came when Gerald Durrell arrived at Karanambu in 1950 to catch animals for zoos in England and Europe [9]. Together, he and Tiny caught capybaras and large tortoises. They also managed to chase down a giant anteater on the savanna (Figure 90).

After McTurk had lassoed the anteater, he and Durrell dismounted and wrestled the great beast to the ground. This was insanely brave, especially in view of the anteater's wicked claws. A persistent tale is still told in the Guyana hinterland of hunters who discovered the corpses of an anteater and a jaguar that had died like Sherlock Holmes and Professor Moriarty, locked together in a final deadly embrace.

Figure 91 Diane McTurk and her most recent pair of otter orphans, Philip (left) and Belle. Both of these giant otters, *Pteronura brasiliensis*, have now been released successfully into the wild, although Philip still turns up occasionally, expecting (and getting) a fish feast from the indulgent workers at the lodge.

More recently Karanambu has been in the news again, thanks to the conservation efforts of Tiny McTurk's daughter Diane. Over the years she has hand-raised about fifty orphaned giant river otters, releasing many of them back into the wild (Figure 91) [10, 11].

Diane's joy in helping to save Guyana's endangered species is infectious. It has spread to nearby Caiman House in the Makushi village of Yupukari, where many of the villagers work at a variety of ecological projects. The Makushi Caiman Research Project has weighed and measured almost a thousand caimans, and has followed the fates of hundreds of nestlings [12].

We are able to quantify many aspects of the world's ecosystems and of our interactions with them. We can use these measurements to track a course towards saving and re-establishing the green equilibria on which we all depend. But we cannot quantify the joy in this enterprise that is shared by the ecologists who study these ecosystems and by the people who live in and depend on them. It is a never-ending source of joy to see how quickly the resilient webs of life can respond to their efforts. And it is joy's incalculable effect that gives me the greatest hope for our species and for our planet.

L'Envoi

Among the ruined temples of Ayutthaya, the ancient capital of the Kingdom of Siam, there is a temple called Wat Mahathat where a banyan tree has overgrown a statue of the Buddha (Figure 92). The tree's roots caress and twine around the Buddha's face.

For most of the history of our planet, except during brief periods of geological upheaval, green equilibria have been the norm. Paleontologists argue over the exact trajectory of increases in species richness and diversity that have taken place since the beginning of large multicellular life more than half a billion years ago [1, 2]. But it is likely that we are living in one of the most biologically diverse periods of Earth's history.

Figure 92 Humankind and nature are symbolically woven together in this overgrown statue of the Buddha in Thailand's old capital of Ayatthura.

During most of the amazing story of this history of life, there were no intelligent observers present to be amazed. Now our species has evolved to become both observers and participants. We have learned to understand the planet. And it is our task, like the Buddha cradled by the banyan, to become a seamless part of it.

Endnotes

Chapter 1 Notes

1. There are some highly unusual ecosystems near volcanic vents in the deep ocean, far from sunlight. Bacteria in these ecosystems take the place of plants, getting their energy from hydrogen sulfide from the vents. The tube worms and other animals living in these ecosystems depend in turn on the bacteria. But all these animals evolved from ancestors that had originally inhabited ecosystems near the Earth's surface. Those surface ecosystems, of course, were driven by sunlight.

2. They were able to get away with this because Florida's environmental protection regulations were even more lax forty years ago than they currently are.

Chapter 2 Notes

1. Only the *P. dardanus* females mimic the various model species. The males exhibit a single non-mimetic pattern. The difference can be traced to differences in behavior between the sexes. Females must remain still while they are courted by males. This inactivity makes them sitting ducks (or sitting butterflies) for predators. The males, on the other hand, are constantly flitting about, guarding their territories, and competing with other males for the attention of the females. The males are harder for predatory birds to catch. While *Papilio* males do not seem to care about the appearance of the females they are courting, *Papilio* females have been shown to prefer males of the non-mimetic pattern [7]. Thus, if a male were to imitate a model species, he might be more likely to survive predation, but females would not recognize him and his genes would soon disappear! In the *Papilio* males, the imperative of sexual selection, in this case the ability to be recognized by females, trumps any protection that might be conferred by mimicry.

2. The term parataxonomist was introduced by ecologist Daniel Janzen in 1993 to describe local people who have been recruited to identify animal and plant species. Some of New Guinea's parataxonomists have published papers and have gone on to advanced degrees.

Chapter 3 Notes

1. The simple wooden boats that are used by the fishermen are well-adapted to the local conditions. Their twin outriggers keep them stable in the rough seas around Anilao. And they are easily repaired, as we discovered when we were caught in a sudden squall. Our boat was smashed against the rocks, bending the propeller shaft. The captain and his assistant waded ashore with the shaft and pounded it straight using a rock as an anvil and

a block of driftwood as a hammer. Such resourcefulness, gleaned from an earlier and much more challenging life as fishermen, aids these boatmen as they make the transition from a fisheries-based to a tourist economy.

Chapter 4 Notes

1. Not all is smooth sailing. As I wrote this, a SCUBA diver was cited for poaching 47 undersized spiny lobsters off the waters of Laguna Beach. This area had become part of the California coastal reserve system only a month earlier.
2. We found that the Santa Cruz population of *Drosophila pseudoobscura*, like other populations of this fly on the mainland, had indeed undergone genetic changes over the decades. But whether these changes were the result of human-caused alterations in the environment or would have taken place anyway remains unresolved [9].
3. Henry tragically committed suicide in 1985.
4. Others may quibble that the title of most oxymoronic should go to jumbo shrimp, but I disagree. The jumbo shrimp is simply an effort by the advertising industry to sell more shrimp, while the pygmy mammoth is recognized as a distinct species, *Mammuthus exilis*.
5. The most ancient fox remains that have yet been found on the islands are only about six thousand years old, so there is a possibility that humans brought them from the mainland and they evolved to their small size during a span of just a few thousand years [20].

Chapter 5 Notes

1. *Tepui* means house of the gods, in the language of the Pemon tribe who live on the savannahs to the south of these tablelands. The precipitous *tepuis* were the inspiration for Arthur Conan Doyle's *Lost World*, in which he imagined them to be cloud-shrouded eyries inhabited by snarling dinosaurs.
2. This rate of movement may seem quite poky, but it depends on your time-scale. Four centimeters a year adds up to eight hundred kilometers in a mere twenty million years.

Chapter 8 Notes

1. Mammals on islands have not always gotten smaller. Recently-discovered fossil remains of giant rabbits of the island of Minorca, *Nuralagus rex*, show that they were six times as large as modern rabbits. Because these super-rabbits had small ears and were unable to leap, they were probably not threatened by predators. The rabbits occupied a low trophic level, which means there was abundant food for them on the island. This food abundance probably allowed them to grow larger as a result of sexual selection for big males that could compete for females [2].
2. These data were some of the results from a spectacular scientific contest, in which the immense resources of the NIH were pitted against Venter's private company Celera in a race to sequence the first human genome. The NIH was hoping to prevent Celera from patenting large numbers of human genes. Both sides claimed victory in the race, but the Clinton Administration intervened in a Solomonic fashion in 2000 and got them to agree to share the credit [9].

Chapter 10 Notes

1. At the present time, much of New Guinea's politics is based on conflicts between languages or language groups. These are called *wantoks* ("one-talks") in Papua New Guinea's recently-emerged pidgin language *tok pisin*. All members of a *wantok* are expected to help each other, even to the extent of giving up everything that they own to help other members of the group. *Wantoks* can work after a fashion in a village setting, but they have slowed the emergence of an urban economy [17].

2. All is still not sweetness and light even at the sing-sings. Alcohol is banned during the festivities, but tribal animosities surface during the drunken evening celebrations that follow. This often leads to bloodshed. When we attended the sing-sing at the village of Tufi on the eastern end of the island, we learned that the organizers were trying to reduce the occasions for conflict by giving a first prize to every one of the contestants in the sing-sing's beauty contest.

3. The last of the moa hunts may have overlapped with the first European explorers to arrive in New Zealand. J.W. Hamilton, private secretary to Governor Fitzroy (yes, that Fitzroy, the former commander of the brig *Beagle* on which Darwin sailed), recounts interviews in the 1840s with three old Maoris who had seen and hunted moas. One of the Maoris said that he had also met Captain Cook when Cook first visited New Zealand in 1769 [24].

4. The Inuit and other peoples of the far north have more recent Asiatic origins.

Chapter 11 Notes

1. Of course, lots of other things have happened to human brains, even to some of their most ancient components. The white matter that lies beneath our cerebral cortex "wires" different parts of the brain together. This part of the brain has increased greatly in complexity in the human lineage. And the ancient cerebellum, the "little brain" that nestles close to our brain stem and enables us to maintain our balance, has become more versatile in humans and is involved in a variety of functions including language [21].

2. Several kinds of RNA molecules can also play roles in gene regulation.

3. Such between-species comparisons are tricky. It is essential to confine the analysis to bits of genes that have exactly the same sequence in all of the different species that are being compared. Even with this restriction, it is still possible to compare RNA production from more than half of the genes that are shared by humans, chimpanzees, and macaques.

4. See Figure 64 for some examples of such sites in which different bases are found in different humans. These sites are typical SNPs.

Chapter 12 Notes

1. Nothing unusual in that, American readers might say, but remember that in Papua New Guinea people are supposed to drive on the left.

Bibliography

Introduction

1. Fyumagwa, R.D., et al., *Ecology and control of ticks as disease vectors in wildlife of the Ngorongoro Crater, Tanzania.* South African Journal of Wildlife Research, 2007. **37**(1): p. 79–90.
2. Ramankutty, N., et al., *The global distribution of cultivable lands: current patterns and sensitivity to possible climate change.* Global Ecology and Biogeography, 2002. **11**(5): p. 377–92.
3. United Nations, *World Population Prospects: The 2002 Revision, Vol. I, Comprehensive Tables,* 2002, United Nations Publications: New York.
4. Pinker, S., *The Better Angels of Our Nature: Why Violence Has Declined,*2011. Viking Adult: New York.

Chapter One

1. Odum, E.P., *Fundamentals of Ecology,* 3rd ed., 1971. Saunders: Philadelphia.
2. Snow, D.R., *The first Americans and the differentiation of hunter-gatherer cultures,* in *The Cambridge History of the Native Peoples of the Americas,* B.G. Trigger and W.E. Washburn, Editors, 1996. Cambridge University Press: Cambridge. p. 125–99.
3. Randall, J.M., M. Rejmánek, and J.C. Hunter, *Characteristics of the exotic flora of California.* Fremontia, 1998. **26**: p. 3–12.
4. Nefstead, P., *History of Rancho Laguna,* 2012. Available from: http://www.sedgwick.org/na/families/robert1613/B/4/7/4/2/3/RanchoLaLagunaHistory.html
5. Hendry, G.W., and M.P. Kelly, *The plant content of adobe bricks: with a note on adobe brick making.* California Historical Society Quarterly, 1925. **4**(4): p. 361–73.
6. Hendry, G.W., *The adobe brick as a historical source: reporting further studies in adobe brick analysis.* Agricultural History, 1931. **5**(3): p. 110–27.
7. Sharma, M.P., and W.H. Vanden Born, *The biology of Canadian weeds. 27. Avena fatua L.* Canadian Journal of Plant Science, 1978. **58**: p. 141–57.
8. Hutchinson, G.E., *Homage to Santa Rosalia or why are there so many kinds of animals?* The American Naturalist, 1959. **93**(870): p. 145–59.
9. Paine, C.E.T., and H. Beck, *Seed predation by neotropical rain forest mammals increases diversity in seedling recruitment.* Ecology, 2007. **88**(12): p. 3076–87.
10. Punzalan, D., F.H. Rodd, and K.A. Hughes, *Perceptual processes and the maintenance of polymorphism through frequency-dependent predation.* Evolutionary Ecology, 2005. **19**(3): p. 303–20.
11. MacArthur, R.H., and E.O. Wilson, *The Theory of Island Biogeography,* 1967. Princeton University Press: Princeton NJ.

251

12. Simberloff, D.S., and E.O. Wilson, *Experimental zoogeography of islands—A 2-year record of colonization.* Ecology, 1970. **51**(5): p. 934–7.

13. Hawkins, C.C., W.E. Grant, and M.T. Longnecker, *Effect of subsidized house cats on California birds and rodents.* Transactions of the Western Section of The Wildlife Society, 1999. **35**: p. 29–33.

14. Stevens, M., C.J. Hardman, and C.L. Stubbins, *Conspicuousness, not eye mimicry, makes "eyespots" effective antipredator signals.* Behavioral Ecology, 2008. **19**(3): p. 525–31.

15. Rull, V., *Is the "Lost World" really lost? Palaeoecological insights into the origin of the peculiar flora of the Guayana Highlands.* Naturwissenschaften, 2004. **91**(3): p. 139–42.

16. Kok, P.J.R., et al., *A new species of Colostethus (Anura: Dendrobatidae) with maternal care from Kaieteur National Park, Guyana.* Zootaxa, 2006. (1238): p. 35–61.

17. Grinnell, J., *The niche-relationships of the California Thrasher.* The Auk, 1917. **34**(4): p. 427–33.

18. Elton, C.S., *Animal Ecology,* 1927. Sidgwick & Jackson: London.

19. Bourne, G.R., et al., *Vocal communication and reproductive behavior of the frog Colostethus beebei in Guyana.* Journal of Herpetology, 2001. **35**(2): p. 272–81.

20. Romero, G.Q., et al., *Nitrogen fluxes from treefrogs to tank epiphytic bromeliads: an isotopic and physiological approach.* Oecologia, 2010. **162**(4): p. 941–9.

21. Giménez, M., *Tenondeté: mitos, leyendas y tradiciones del área Tupi-Guaraní* 1980, Editorial El Foro: Asunciòn.

22. Seymour, R.S., and P.G.D. Matthews, *The role of thermogenesis in the pollination biology of the Amazon waterlily victoria amazonica.* Annals of Botany, 2006. **98**(6): p. 1129–35.

23. Rohde, K., *Latitudinal gradients in species-diversity—The search for the primary cause.* Oikos, 1992. **65**(3): p. 514–27.

24. Mittelbach, G.G., et al., *Evolution and the latitudinal diversity gradient: speciation, extinction and biogeography.* Ecology Letters, 2007. **10**(4): p. 315–31.

25. Gillett, J.B., *Pest pressure, and underestimated factor in evolution, in Taxonomy and Geography, a Symposium,* D. Nichols, editor, 1962. London Systematics Association: London. p. 37–46.

26. Elias, R., et al., *First report of thelaziosis (Thelazia anolabiata) in an Andean Cock of the Rock (Rupicola peruviana) from Peru.* Veterinary Parasitology, 2008. **158**(4): p. 382–3.

27. Valim, M.P., and F.A. Hernandes, *Redescriptions of five species of the feather mite genus Pterodectes Robin, 1877 (Acari: Proctophyllodidae: Pterodectinae), with the proposal of a new genus and a new species.* Acarina, 2008. **16**(2): p. 131–57.

28. Haub, C., *How many people have ever lived on Earth?* 2011. Available from: http://www.prb.org/Articles/2002/HowManyPeopleHaveEverLivedonEarth.aspx.

29. Species Redlist *Anamologlossus beebei.* 2011. Available from: http://www.iucnredlist.org/apps/redlist/details/55052/0.

30. Roff, D.A., and T. Mousseau, *The evolution of the phenotypic covariance matrix: evidence for selection and drift in Melanoplus.* Journal of Evolutionary Biology, 2005. **18**(4): p. 1104–14.

Chapter Two

1. Bond, A.B., and A.C. Kamil, *Visual predators select for crypticity and polymorphism in virtual prey.* Nature, 2002. **415**(6872): p. 609–13.

2. Beukers-Stewart, B., J. Beukers-Stewart, and G. Jones, *Behavioural and developmental responses of predatory coral reef fish to variation in the abundance of prey*. Coral Reefs, 2011. **30**(3): p. 855–64.

3. Ming Chou, L., et al., *Temporal changes in reef community structure at Bintan Island (Indonesia) suggest need for integrated management*. Pacific Science, 2010. **64**(1): p. 99–111.

4. Poulton, E.B., *Notes upon, or suggested by, the colours, markings and protective attitudes of certain lepidopterous larvae and pupae, and of a phytophagous hymenopterous larva*. Transactions of the Entomological Society of London 1884. p. 27–60.

5. Poulton, E.B., *Further experiments upon the colour-relation between certain lepidopterous larvae, pupae, cocoons, and imagines and their surroundings*. Transactions of the Entomological Society of London, 1892. **40**: p. 293–487.

6. Noor, M.A.F., R.S. Parnell, and B.S. Grant, *A reversible color polyphenism in American Peppered Moth (Biston betularia cognataria) caterpillars*. PLoS ONE, 2008. **3**(9), e3142.

7. Krebs, R.A., and D.A. West, *Female mate preference and the evolution of female-limited Batesian mimicry*. Evolution, 1988. **42**(5): p. 1101–4.

8. Clarke, C.A., and P.M. Sheppard, *The genetics of Papilio dardanus, Brown. Iii. Race antinorii from Abyssinia and race meriones from Madagascar*. Genetics, 1960. **45**(6): p. 683–98.

9. Clark, R., et al., *Colour pattern specification in the Mocker swallowtail Papilio dardanus: the transcription factor invected is a candidate for the mimicry locus H*. Proceedings of the Royal Society B: Biological Sciences, 2008. **275**(1639): p. 1181–8.

10. Janzen, D.H., *Herbivores and the number of tree species in tropical forests*. American Naturalist, 1970. **104**: p. 501–29.

11. Connell, J.H., *On the role of natural enemies in preventing competitive exclusion in some marine animals and in rain forest trees*, in *Dynamics of Populations*, P.J. Den Boer and G. Gradwell, editors, 1971. PUDOC: New York.

12. Wallace, A.R., *Tropical Nature and Other Essays*, 1878. Macmillan: London.

13. Condit, R., S.P. Hubbell, and R.B. Foster, *Recruitment near conspecific adults and the maintenance of tree and shrub diversity in a neotropical forest*. American Naturalist, 1992. **140**(2): p. 261–86.

14. Wills, C., et al., *Strong density- and diversity-related effects help to maintain tree species diversity in a neotropical forest*. Proceedings of the National Academy of Sciences, 1997. **94**(4): p. 1252–7.

15. Wills, C., et al., *Nonrandom processes maintain diversity in tropical forests*. Science, 2006. **311**(5760): p. 527–31.

16. Packer, A., and K. Clay, *Development of negative feedback during successive growth cycles of black cherry*. Proceedings of the Royal Society B: Biological Sciences, 2004. **271**(1536): p. 317–24.

17. Marhaver, K.L., et al., *Janzen–Connell effects in a broadcast-spawning Caribbean coral: Distance-dependent survival of larvae and settlers*. Ecology. http://dx.doi.org/10.1890/12-0985.1 (in advance of publication).

18. Novotny, V., et al., *Guild-specific patterns of species richness and host specialization in plant-herbivore food webs from a tropical forest*. Journal of Animal Ecology, 2010. **79**(6): p. 1193–203.

19. Novotny, V., *Notebooks from New Guinea: Field Notes of a Tropical Biologist*, 2009. Oxford University Press: Oxford.

20. Kost, C., M. Tremmel, and R. Wirth, *Do leaf cutting ants cut undetected? Testing the effect of ant-induced plant defences on foraging decisions in Atta colombica.* PLoS ONE, 2011. **6**(7): p. e22340.

21. Matsuki, M., W.J. Foley, and R.B. Floyd, *Role of volatile and non-volatile plant secondary metabolites in host tree selection by Christmas beetles.* Journal of Chemical Ecology, 2011. **37**(3): p. 286–300.

22. Bar-Yam, S., and D.H. Morse, *Host–plant choice behavior at multiple life-cycle stages: the roles of mobility and early growth in decision-making.* Ethology, 2011. **117**(6): p. 508–19.

23. Xue, M., et al., *Effects of four host plants on biology and food utilization of the cutworm, Spodoptera litura.* Journal of Insect Science (Tucson), 2010. **10**: p. 22.

24. Mugrabi-Oliveira, E., and G.R.P. Moreira, *Conspecific mimics and low host plant availability reduce egg laying by Heliconius erato phyllis (Fabricius) (Lepidoptera: Nymphalidae).* Revista Brasileira de Zoologia, 1996. **13**(4): p. 929–37.

25. Patricelli, D., et al., *To lay or not to lay: oviposition of Maculinea arion in relation to Myrmica ant presence and host plant phenology.* Animal Behaviour, 2011. **82**(4): p. 791–9.

26. Erwin, T.L., *Tropical forests: their richness in Coleoptera and other arthropod species.* Coleopterists Bulletin, 1982. **36**(1): p. 74–5.

27. Novotny, V., et al., *Low host specificity of herbivorous insects in a tropical forest.* Nature, 2002. **416**(6883): p. 841–4.

Chapter Three

1. Hares, M., *Forest conflict in Thailand: Northern minorities in focus.* Environmental Management, 2009. **43**(3): p. 381–95.

2. Srikosamatara, S., *Density and biomass of large herbivores and other mammals in a dry tropical forest, western Thailand.* Journal of Tropical Ecology, 1993. **9**(1): p. 33–43.

3. Karanth, K.U., and M.E. Sunquist, *Population structure, density and biomass of large herbivores in the tropical forests of Nagarahole, India.* Journal of Tropical Ecology, 1992. **8**(01): p. 21–35.

4. PAD Review Team, *Thailand: National Report on Protected Areas and Development*, 2003. ICEM: Indooroopilly, Queensland.

5. Simcharoen, S., et al., *How many tigers Panthera tigris are there in Huai Kha Khaeng Wildlife Sanctuary, Thailand? An estimate using photographic capture-recapture sampling.* Oryx, 2007. **41**(04): p. 447–53.

6. Walston, J., et al., *Bringing the tiger back from the brink-the six percent solution.* PLoS Biology, 2010. **8**(9), e1000485.

7. BBC News *India wild tiger census shows population rise.* 2011. Available from: http://www.bbc.co.uk/news/world-south-asia-12877560.

8. McShea, W.J., et al., *Finding a needle in the haystack: Regional analysis of suitable Eld's deer (Cervus eldi) forest in Southeast Asia.* Biological Conservation, 2005. **125**(1): p. 101–11.

9. Poonswad, P., et al., *Comparison of cavity modification and community involvement as strategies for hornbill conservation in Thailand.* Biological Conservation, 2005. **122**(3): p. 385–93.

10. Russ, G.R., and A.C. Alcala, *Marine reserves: Rates and patterns of recovery and decline of large predatory fish.* Ecological Applications, 1996. **6**(3): p. 947–61.

11. Manuel, J., *Apo Reef after Typhoon Caloy*. 2006. Available from: http://newsfromkkp.blog-spirit.com/archive/2006/06/03/apo-reef-after-typhoon-caloy1.html.

12. White, A.T., and H.P. Vogt, *Philippine coral reefs under threat: Lessons learned after 25 years of community-based reef conservation*. Marine Pollution Bulletin, 2000. **40**(6): p. 537–50.

13. White, A.T., H.P. Vogt, and T. Arin, *Philippine coral reefs under threat: The economic losses caused by reef destruction*. Marine Pollution Bulletin, 2000. **40**(7): p. 598–605.

14. Haring, E., et al., *Convergent evolution and paraphyly of the hawk-eagles of the genus Spizaetus (Aves, Accipitridae)—phylogenetic analyses based on mitochondrial markers*. Journal of Zoological Systematics and Evolutionary Research, 2007. **45**(4): p. 353–65.

15. Margolis, M., *Treasuring the Pantanal*. International Wildlife, 1995. **25**: p. 12–21.

Chapter Four

1. Hill, K.T., et al., *Assessment Of The Pacific Sardine (Sardinops sagax caerulea) Population For U.S. Management In 2006*, NOAA, 2006. Southwest Fisheries Science Center: La Jolla.

2. Halpern, B.S., *Strong top-down control in southern California kelp forest ecosystems*. Science, 2006. **312**(5777): p. 1230–2.

3. Paddack, M.J., and J.A. Estes, *Kelp forest fish populations in marine reserves and adjacent exploited areas of central California*. Ecological Applications, 2000. **10**(3): p. 855–70.

4. Stevenson, C., et al., *High apex predator biomass on remote Pacific islands*. Coral Reefs, 2007. **26**(1): p. 47–51.

5. DeMartini, E.E., et al., *Differences in fish-assemblage structure between fished and unfished atolls in the northern Line Islands, central Pacific*. Marine Ecology-Progress Series, 2008. **365**: p. 199–215.

6. Ruttenberg, B.I., et al., *Predator-induced demographic shifts in coral reef fish assemblages*. PLoS ONE, 2011. **6**(6), e21062.

7. Kay, M.C., et al., *Collaborative assessment of California spiny lobster population and fishery responses to a marine reserve network*. Ecological Applications, 2012. **22**(1): p. 322–35.

8. Krkosek, M., et al., *Effects of parasites from salmon farms on productivity of wild salmon*. Proceedings of the National Academy of Sciences of the United States of America, 2011. **108**(35): p. 14700–4.

9. Dobzhansky, T., et al., *Genetics of natural populations. XXXV. A progress report on genetic changes in populations of Drosophila pseudoobscura in the American southwest*. Evolution, 1964. **18**(2): p. 164–76.

10. McNab, B.K., *Energy-conservation and the evolution of flightlessness in birds*. American Naturalist, 1994. **144**(4): p. 628–42.

11. Erlandson, J.M., et al., *Paleoindian seafaring, maritime technologies, and coastal foraging on California's Channel Islands*. Science, 2011. **331**(6021): p. 1181–5.

12. Rick, T.C., et al., *Stable isotope analysis of dog, fox, and human diets at a Late Holocene Chumash village (CA-SRI-2) on Santa Rosa Island, California*. Journal of Archaeological Science, 2011. **38**(6): p. 1385–93.

13. Jones, T.L., et al., *The protracted Holocene extinction of California's flightless sea duck (Chendytes lawi) and its implications for the Pleistocene overkill hypothesis*. Proceedings of the National Academy of Sciences, 2008. **105**(11): p. 4105–8.

14. Vellanoweth, R.L., et al., *A 6,000 year old red abalone midden from Otter Point, San Miguel Island, California*. North American Archaeologist, 2006. **27**(1): p. 69–90.

15. Braje, T.J., et al., *Fishing from past to present: continuity and resilience of red abalone fisheries on the Channel Islands, California*. Ecological Applications, 2009. **19**(4): p. 906–19.

16. Walter, H.S., and L.A. Taha, *Regeneration of bishop pine (Pinus muricata) in the absence and presence of fire: a case study from Santa Cruz Island, California*, in *Fifth California Islands Symposium*, 1999. Santa Barbara Museum of Natural History: Santa Barbara. p. 172–81.

17. Parkes, J.P., et al., *Rapid eradication of feral pigs (Sus scrofa) from Santa Cruz Island, California*. Biological Conservation, 2010. **143**(3): p. 634–41.

18. Murphy, D.D., and B.A. Wilcox, *Butterfly diversity in natural habitat fragments: a test of the validity of vertebrate-based management*, in *Wildlife 2000: Modeling habitat relationships of terrestrial vertebrates*, J. Verner, M.L. Morrison, and C.J. Ralph, editors, 1986. University of Wisconsin Press: Madison WI. p. 287–92.

19. Roemer, G.W., *Golden eagles, feral pigs, and insular carnivores: How exotic species turn native predators into prey*. Proceedings of the National Academy of Sciences, 2001. **99**(2): p. 791–6.

20. Rick, T.C., et al., *Origins and antiquity of the island fox (Urocyon littoralis) on California's Channel Islands*. Quaternary Research, 2009. **71**(2): p. 93–8.

21. Rick, T.C., et al., *Dogs, humans and island ecosystems: the distribution, antiquity and ecology of domestic dogs (Canis familiaris) on California's Channel Islands, USA*. Holocene, 2008. **18**(7): p. 1077–87.

22. Anlauf, H., L. D'Croz, and A. O'Dea, *A corrosive concoction: The combined effects of ocean warming and acidification on the early growth of a stony coral are multiplicative*. Journal of Experimental Marine Biology and Ecology, 2011. **397**(1): p. 13–20.

23. Wild, C., et al., *Climate change impedes scleractinian corals as primary reef ecosystem engineers*. Marine and Freshwater Research, 2011. **62**(2): p. 205–15.

24. Pandolfi, J.M., et al., *Projecting coral reef futures under global warming and ocean acidification*. Science (Washington DC), 2011. **333**(6041): p. 418–22.

25. Cuif, J.-P., and Y. Dauphin, *The two-step mode of growth in the scleractinian coral skeletons from the micrometre to the overall scale*. Journal of Structural Biology, 2005. **150**(3): p. 319–31.

26. De'ath, G., J.M. Lough, and K.E. Fabricius, *Declining coral calcification on the Great Barrier Reef*. Science, 2009. **323**(5910): p. 116–19.

27. Fabricius, K.E., et al., *Losers and winners in coral reefs acclimatized to elevated carbon dioxide concentrations*. Nature Climate Change, 2011. **1**(3): p. 165–9.

28. Reynaud, S., et al., *Interacting effects of CO_2 partial pressure and temperature on photosynthesis and calcification in a scleractinian coral*. Global Change Biology, 2003. **9**(11): p. 1660–8.

29. Fine, M., and D. Tchernov, *Scleractinian coral species survive and recover from decalcification*. Science, 2007. **315**(5820): p. 1811.

30. Scheibner, C., and R.P. Speijer, *Late Paleocene–early Eocene Tethyan carbonate platform evolution. A response to long- and short-term paleoclimatic change*. Earth-Science Reviews, 2008. **90**(3–4): p. 71–102.

31. Iglesias-Rodriguez, M.D., et al., *Phytoplankton calcification in a high-CO_2 world*. Science, 2008. **320**(5874): p. 336–40.

32. Venn, A., et al., *Live tissue imaging shows reef corals elevate pH under their calcifying tissue relative to seawater.* PLoS ONE, 2011. **6**(5), e20013.

Chapter Five

1. Graham, J.B., and H.J. Lee, *Breathing air in air: In what ways might extant amphibious fish biology relate to prevailing concepts about early tetrapods, the evolution of vertebrate air breathing, and the vertebrate land transition?* Physiological and Biochemical Zoology, 2004. **77**: p. 720–31.

2. Brauner, C.J., et al., *Transition in organ function during the evolution of air breathing: insights from Arapaima gigas, an obligate airbreathing teleost from the Amazon.* Journal of Experimental Biology, 2004. **207**: p. 1433–8.

3. Schulte, P., et al., *The Chicxulub asteroid impact and mass extinction at the Cretaceous-Paleogene boundary.* Science, 2010. **327**(5970): p. 1214–18.

4. Springer, M.S., et al., *Placental mammal diversification and the Cretaceous–Tertiary boundary.* Proceedings of the National Academy of Sciences, 2003. **100**(3): p. 1056–61.

5. Cooper, A., and D. Penny, *Mass survival of birds across the Cretaceous–Tertiary boundary: Molecular evidence.* Science, 1997. **275**(5303): p. 1109–13.

6. Keyte, A.L., and K.K. Smith, *Developmental origins of precocial forelimbs in marsupial neonates.* Development, 2010. **137**(24): p. 4283–94.

7. Simpson, G.G., *Splendid Isolation: The Curious History of South American Mammals*, 1980. Yale University Press: New Haven.

8. Keynes, R.D., ed., *Charles Darwin's Beagle Diary*, 1988. Cambridge University Press: Cambridge.

9. Woodburne, M., *The great American biotic interchange: Dispersals, tectonics, climate, sea level and holding pens.* Journal of Mammalian Evolution, 2010. **17**(4): p. 245–64.

10. Burnham, R.J., and A. Graham, *The history of neotropical vegetation: New developments and status.* Annals of the Missouri Botanical Garden, 1999. **86**(2): p. 546–89.

11. Chen, Z.-Q., and M.J. Benton, *The timing and pattern of biotic recovery following the end-Permian mass extinction.* Nature Geoscience, 2012. **5**(6): p. 375–83.

12. Woodhead, J., et al., *The big crunch: Physical and chemical expressions of arc/continent collision in the Western Bismarck arc.* Journal of Volcanology and Geothermal Research, 2010. **190**(1–2): p. 11–24.

13. Deiner, K., et al., *A passerine bird's evolution corroborates the geologic history of the island of New Guinea.* PLoS ONE, 2011. **6**(5), e19479.

14. Wroe, S., *Cranial mechanics compared in extinct marsupial and extant African lions using a finite-element approach.* Journal of Zoology, 2008. **274**(4): p. 332–9.

15. Field, J., M. Fillios, and S. Wroe, *Chronological overlap between humans and megafauna in Sahul (Pleistocene Australia, New Guinea): A review of the evidence.* Earth-Science Reviews, 2008. **89**(3–4): p. 97–115.

16. Irestedt, M., et al., *An unexpectedly long history of sexual selection in birds-of-paradise.* BMC Evolutionary Biology, 2009. **9**, article 235.

17. Lovari, S., et al., *Restoring a keystone predator may endanger a prey species in a human-altered ecosystem: the return of the snow leopard to Sagarmatha National Park.* Animal Conservation, 2009. **12**(6): p. 559–70.

18. Cande, S.C., and D.R. Stegman, *Indian and African plate motions driven by the push force of the Réunion plume head*. Nature, 2011. **475**(7354): p. 47–52.

19. Thapar, V., *Land of the Tiger: A Natural History of the Indian Subcontinent*, 1998. University of California Press: Berkeley.

20. Visser, J.N.J., *Geography and climatology of the late Carboniferous to Jurassic Karoo Basin in south-western Gondwana*. Annals of the South African Museum, 1991. **99**(12): p. 415–32.

21. Aitchison, J.C., J.R. Ali, and A.M. Davis, *When and where did India and Asia collide?* Journal of Geophysical Research-Solid Earth, 2007. **112**(B5), article B05423.

22. Zhisheng, A., et al., *Evolution of Asian monsoons and phased uplift of the Himalayan Tibetan plateau since Late Miocene times*. Nature, 2001. **411**(6833): p. 62–6.

23. Raymo, M.E., and W.F. Ruddiman, *Tectonic forcing of late Cenozoic climate*. Nature, 1992. **359**(6391): p. 117–22.

Chapter Six

1. Patnaik, R., and P. Chauhan, *India at the cross-roads of human evolution*. Journal of Biosciences, 2009. **34**(5): p. 729–47.

2. Grove, M., *Change and variability in Plio-Pleistocene climates: modelling the hominin response*. Journal of Archaeological Science, 2011. **38**(11): p. 3038–47.

3. Wood, B., *Reconstructing human evolution: Achievements, challenges, and opportunities*. Proceedings of the National Academy of Sciences, 2010. **107**(Supplement 2): p. 8902–9.

4. Lovejoy, C.O., *Reexamining human origins in light of Ardipithecus ramidus*. Science, 2010. **327**(5967): p. 781.

5. Patterson, N., et al., *Genetic evidence for complex speciation of humans and chimpanzees*. Nature, 2006. **441**(7097): p. 1103–8.

6. Parent, C.E., A. Caccone, and K. Petren, *Colonization and diversification of Galapagos terrestrial fauna: a phylogenetic and biogeographical synthesis*. Philosophical Transactions of the Royal Society B – Biological Sciences, 2008. **363**(1508): p. 3347–61.

7. Ndithia, H., M.R. Perrin, and M. Waltert, *Breeding biology and nest site characteristics of the Rosy-faced Lovebird Agapornis roseicollis in Namibia*. Ostrich, 2007. **78**(1): p. 13–20.

8. Eberhard, J.R., *Evolution of nest-building behavior in Agapornis parrots*. Auk, 1998. **115**(2): p. 455–64.

9. Dilger, W.C., *The behavior of lovebirds*. Scientific American, 1961. **206**: p. 88–98.

10. Maan, M.E., and O. Seehausen, *Ecology, sexual selection and speciation*. Ecology Letters, 2011. **14**(6): p. 591–602.

11. Orr, M.R., and T.B. Smith, *Ecology and speciation*. Trends in Ecology & Evolution, 1998. **13**(12): p. 502–6.

12. Rieseberg, L.H., and B.K. Blackman, *Speciation genes in plants*. Annals of Botany, 2010. **106**(3): p. 439–55.

13. Kulathinal, R.J., and R.S. Singh, *The molecular basis of speciation: from patterns to processes, rules to mechanisms*. Journal of Genetics, 2008. **87**(4): p. 327–38.

14. Gavrilets, S., *Perspective: Models of speciation: What have we learned in 40 years?* Evolution, 2003. **57**(10): p. 2197–215.

15. Coyne, J.A., and H.A. Orr, *Patterns of speciation in Drosophila.* Evolution, 1989. **43**(2): p. 362–81.

16. Yamamichi, M., J. Gojobori, and H. Innan, *An autosomal analysis gives no genetic evidence for complex speciation of humans and chimpanzees.* Molecular Biology and Evolution, 2012. **29**(1): p. 145–56.

17. Becquet, C., and M. Przeworski, *Learning about modes of speciation by computational approaches.* Evolution, 2009. **63**(10): p. 2547–62.

18. Locke, D.P., et al., *Comparative and demographic analysis of orang-utan genomes.* Nature, 2011. **469**(7331): p. 529–33.

19. Presgraves, D.C., *Sex chromosomes and speciation in Drosophila.* Trends in Genetics, 2008. **24**(7): p. 336–43.

Chapter Seven

1. Yokoyama, Y., et al., *Gamma-ray spectrometric dating of late Homo erectus skulls from Ngandong and Sambungmacan, Central Java, Indonesia.* Journal of Human Evolution, 2008. **55**(2): p. 274–7.

2. Brumm, A., et al., *Stone technology at the Middle Pleistocene site of Mata Menge, Flores, Indonesia.* Journal of Archaeological Science, 2010. **37**(3): p. 451–73.

3. Kaifu, Y., et al., *Craniofacial morphology of Homo floresiensis: Description, taxonomic affinities, and evolutionary implication.* Journal of Human Evolution, 2011. **61**(6): p. 644–82.

4. Martinon-Torres, M., et al., *Dental remains from Dmanisi (Republic of Georgia): Morphological analysis and comparative study.* Journal of Human Evolution, 2008. **55**(2): p. 249–73.

5. Ferring, R., et al., *Earliest human occupations at Dmanisi (Georgian Caucasus) dated to 1.85–1.78 Ma.* Proceedings of the National Academy of Sciences of the United States of America, 2011. **108**(26): p. 10432–6.

6. Ron, H., and S. Levi, *When did hominids first leave Africa?: New high-resolution magnetostratigraphy from the Erk-el-Ahmar Formation, Israel.* Geology, 2001. **29**(10): p. 887–90.

7. Zaim, Y., et al., *New 1.5 million-year-old Homo erectus maxilla from Sangiran (Central Java, Indonesia).* Journal of Human Evolution, 2011. **61**(4): p. 363–76.

8. Indriati, E., et al., *The age of the 20 meter Solo River Terrace, Java, Indonesia and the survival of Homo erectus in Asia.* PLoS ONE, 2011. **6**(6): p. e21562.

9. van den Bergh, G.D., et al., *The Liang Bua faunal remains: a 95 k.yr. sequence from Flores, East Indonesia.* Journal of Human Evolution, 2009. **57**(5): p. 527–37.

10. Anantharaman, S., D.C. Dassarma, and P.A. Kumar, *A new species of Quaternary hippopotamid from Bhima Valley, Karnataka.* Journal of the Geological Society of India, 2005. **66**(2): p. 209–16.

11. Mounier, A., F. Marchal, and S. Condemi, *Is Homo heidelbergensis a distinct species? New insight on the Mauer mandible.* Journal of Human Evolution, 2009. **56**(3): p. 219–46.

12. Finlayson, C., et al., *Late survival of Neanderthals at the southernmost extreme of Europe.* Nature, 2006. **443**(7113): p. 850–3.

13. Pinhasi, R., et al., *Revised age of late Neanderthal occupation and the end of the Middle Paleolithic in the northern Caucasus.* Proceedings of the National Academy of Sciences, 2011. **108**(21): p. 8611–16.

14. Thieme, H., *Lower Paleolithic hunting spears from Germany.* Nature, 1997. **385:** p. 807–10.

15. Martinón-Torres, M., R. Dennell, and J.M. Bermúdez de Castro, *The Denisova hominin need not be an out of Africa story.* Journal of Human Evolution, 2011. **60**(2): p. 251–5.

16. Garrod, D.A.E., *Excavations in the Caves of the Wadi el-Mughara, 1929 and 1930.* Bulletin of the American Schools of Prehistoric Research, 1931. **7**: p. 5–11.

17. Grun, R., et al., *U-series and ESR analyses of bones and teeth relating to the human burials from Skhul.* Journal of Human Evolution, 2005. **49**(3): p. 316–34.

18. White, T.D., et al., *Pleistocene Homo sapiens from Middle Awash, Ethiopia.* Nature, 2003. **423**(6941): p. 742–7.

19. Cavalier-Smith, T., *Deep phylogeny, ancestral groups and the four ages of life.* Philosophical Transactions of the Royal Society B-Biological Sciences, 2010. **365**(1537): p. 111–32.

20. Macaulay, V., *Single, rapid coastal settlement of Asia revealed by analysis of complete mitochondrial genomes.* Science, 2005. **308** (5724): p. 1034–6.

21. Porath, N., *"They have not progressed enough": Development's negated identities among two indigenous peoples (Orang Asli) in Indonesia and Thailand.* Journal of Southeast Asian Studies, 2010. **41**(2): p. 267–89.

Chapter Eight

1. Pääbo, S., *Molecular cloning of Ancient Egyptian mummy DNA.* Nature, 1985. **314**(644–5).

2. Quintana, J., M. Kohler, and S. Moya-Sola, *Nuralagus rex, gen. et sp. nov., an endemic insular giant rabbit from the Neogene of Minorca (Balearic Islands, Spain).* Journal of Vertebrate Paleontology, 2011. **31**(2): p. 231–40.

3. Krings, M., et al., *Neandertal DNA sequences and the origin of modern humans.* Cell, 1997. **90**(1): p. 19–30.

4. Schmitz, W., et al., *The Neandertal type site revisited: Interdisciplinary investigations of skeletal remains from the Neander Valley, Germany.* Proceedings of the National Academy of Sciences, 2002. **99**(20): p. 13342–7.

5. Teschler-Nicola, M., et al., *No Evidence of Neandertal mtDNA contribution to early modern humans.* PLoS Biology, 2004. **2**(3): p. 313–17.

6. Green, R.E., et al., *A complete Neandertal mitochondrial genome sequence determined by high-throughput sequencing.* Cell, 2008. **134**(3): p. 416–26.

7. Green, R.E., et al., *A draft sequence of the Neandertal genome.* Science, 2010. **328**(5979): p. 710–22.

8. Green, R.E., et al., *Analysis of one million base pairs of Neanderthal DNA.* Nature, 2006. **444**(7117): p. 330–6.

9. Shreeve, J., *The genome war: how Craig Venter tried to capture the code of life and save the world,* 2004. Alfred A. Knopf: New York.

10. Krause, J., et al., *The complete mitochondrial DNA genome of an unknown hominin from southern Siberia.* Nature, 2010. **464**(7290): p. 894–7.

11. Reich, D., et al., *Genetic history of an archaic hominin group from Denisova Cave in Siberia.* Nature, 2010. **468**(7327): p. 1053–60.

12. Gibbons, A., *Who were the Denisovans?* Science, 2011. **333**(6046): p. 1084–7.

13. Rasmussen, M., et al., *An aboriginal Australian genome reveals separate human dispersals into Asia.* Science, 2011. **333**(6052): p. 94–8.

14. Reich, D., et al., *Denisova admixture and the first modern human dispersals into Southeast Asia and Oceania.* American Journal of Human Genetics, 2011. **89**(4): p. 516–28.

15. Lachance, J., et al., *Evolutionary history and adaptation from high-coverage whole-genome sequences of diverse African hunter-gatherers.* Cell, 2012. **150**(3): p. 457–69.

Chapter Nine

1. Elephant News *Tourist attacked by elephant*, 2009. Available from: http://www.elephant-news.com/index.php?id=4875.

2. Waller, D.J., *The Pundits: British exploration of Tibet and Central Asia*, 1990. University Press of Kentucky: Lexington.

3. Galanello, R., and A. Cao, *Alpha-thalassemia.* Genetics in Medicine, 2011. **13**(2): p. 83–8.

4. Lopez, C., et al., *Mechanisms of genetically-based resistance to malaria.* Gene, 2010. **467**(1,2): p. 1–12.

5. Danquah, I., and F.P. Mockenhaupt, *Alpha + -thalassaemia and malarial anaemia.* Trends in Parasitology, 2008. **24**(11): p. 479–81.

6. Modiano, G., et al., *Protection against malaria morbidity—Near-fixation of the alpha-thalassemia gene in a Nepalese population.* American Journal of Human Genetics, 1991. **48**(2): p. 390–7.

7. Fornarino, S., et al., *Mitochondrial and Y-chromosome diversity of the Tharus (Nepal): a reservoir of genetic variation.* BMC Evolutionary Biology, 2009. **9**, article 154.

8. Sakai, Y., et al., *Molecular analysis of alpha-thalassemia in Nepal: correlation with malaria endemicity.* Journal of Human Genetics, 2000. **45**(3): p. 127–32.

9. Ostyn, B., et al., *Incidence of symptomatic and asymptomatic Leishmania donovani Infections in high-endemic foci in India and Nepal: A prospective study.* PLoS Neglected Tropical Diseases, 2011. **5**(10), e1284.

10. Dorji, T., et al., *Diversity and phylogeny of mitochondrial DNA isolated from mithun Bos frontalis located in Bhutan.* Animal Genetics, 2010. **41**(5): p. 554–6.

11. Bollongino, R., et al., *Early history of European domestic cattle as revealed by ancient DNA.* Biology Letters, 2006. **2**(1): p. 155–9.

12. Gerbault, P., et al., *Evolution of lactase persistence: an example of human niche construction.* Philosophical Transactions of the Royal Society B – Biological Sciences, 2011. **366**(1566): p. 863–77.

13. Romero, I.G., et al., *Herders of Indian and European cattle share their predominant allele for lactase persistence.* Molecular Biology and Evolution, 2012. **29**(1): p. 248–59.

14. Burger, J., et al., *Absence of the lactase-persistence-associated allele in early Neolithic Europeans.* Proceedings of the National Academy of Sciences, 2007. **104**(10): p. 3736–41.

15. Keller, A., et al., *New insights into the Tyrolean Iceman's origin and phenotype as inferred by whole-genome sequencing.* Nature Communications, 2012. **3**: p. 698.

16. Beja-Pereira, A., et al., *The origin of European cattle: Evidence from modern and ancient DNA.* Proceedings of the National Academy of Sciences, 2006. **103**(21): p. 8113–18.

17. Bersaglieri, T., et al., *Genetic signatures of strong recent positive selection at the lactase gene.* American Journal of Human Genetics, 2004. **74**(6): p. 1111–20.

18. Sabeti, P.C., et al., *Positive natural selection in the human lineage.* Science, 2006. **312**(5780): p. 1614–20.

19. Hancock, A.M., et al., *Adaptations to new environments in humans: the role of subtle allele frequency shifts.* Philosophical Transactions of the Royal Society B–Biological Sciences, 2010. **365**(1552): p. 2459–68.

20. Moore, L.G., *Human genetic adaptation to high altitude.* High Altitude Medicine & Biology, 2001. **2**(2): p. 257–79.

21. Beall, C.M., *Biodiversity of human populations in mountain environments,* in *Mountain Biodiversity: A Global Assessment,* C. Korner and E.M. Spehn, editors, 2002. Parthenon Publishing Group: New York. p. 199–210.

22. Yi, X., et al., *Sequencing of 50 human exomes reveals adaptation to high altitude.* Science, 2010. **329**(5987): p. 75–8.

23. Beall, C.M., et al., *Natural selection on EPAS1 (HIF2) associated with low hemoglobin concentration in Tibetan highlanders.* Proceedings of the National Academy of Sciences, 2010. **107**(25): p. 11459–64.

24. Simonson, T.S., et al., *Genetic evidence for high-altitude adaptation in Tibet.* Science, 2010. **329**(5987): p. 72–5.

25. Monge-Medrano, C., et al., *La Enfermedad de los Andes (Sindromes Eritemicos),* 1928. Imprenta Americana: Lima.

26. Beall, C.M., *Tibetan and Andean patterns of adaptation to high-altitude hypoxia.* Human Biology, 2000. **72**: p. 201–28.

27. Wills, C., *Rapid recent human evolution and the accumulation of balanced genetic polymorphisms.* High Altitude Medicine and Biology, 2011. **12**: p. 149–55.

28. Hawks, J., et al., *Recent acceleration of human adaptive evolution.* Proceedings of the National Academy of Sciences, 2007. **104**(52): p. 20753–8.

29. Green, R.E., et al., *A draft sequence of the Neandertal genome.* Science, 2010. **328**(5979): p. 710–22.

30. Crisci, J.L., et al., *On characterizing adaptive events unique to modern humans.* Genome Biology and Evolution, 2011. **3**: p. 791–8.

31. Burbano, H.n.A., et al., *Targeted investigation of the Neandertal genome by array-based sequence capture.* Science, 2010. **328**(5979): p. 723–5.

32. Abi-Rached, L., et al., *The shaping of modern human immune systems by multiregional admixture with archaic humans.* Science, 2011. **334**(6052): p. 89–94.

Chapter Ten

1. Leahy, M.J., and M. Crain, *The Land that Time Forgot: Adventures and Discoveries in New Guinea,* 1937. Funk & Wagnalls: New York.

2. Lynch, A.H., et al., *Using the paleorecord to evaluate climate and fire interactions in Australia,* in *Annual Review of Earth and Planetary Sciences,* 2007, Annual Reviews: Palo Alto. p. 215–39.

3. Black, M.P., S.D. Mooney, and V. Attenbrow, *Implications of a 14 200 year contiguous fire record for understanding human-climate relationships at Goochs Swamp, New South Wales, Australia.* Holocene, 2008. **18**(3): p. 437–47.

4. Rule, S., et al., *The aftermath of megafaunal extinction: Ecosystem transformation in Pleistocene Australia.* Science, 2012. **335**(6075): p. 1483–6.

5. Brunskill, G.J., *New Guinea and its coastal seas, a testable model of wet tropical coastal processes: an introduction to Project TROPICS.* Continental Shelf Research, 2004. **24**(19): p. 2273–95.

6. Groube, L., et al., *A 40,000-year-old human occupation site at Huon Peninsula, Papua-New-Guinea.* Nature, 1986. **324**(6096): p. 453–5.

7. Groube, L., *The taming of the rainforests: a model for Late Pleistocene forest exploitation in New Guinea*, in *Foraging and Farming: The Evolution of Plant Exploitation*, D.R. Harris and G.C. Hillman, editors, 1989. Unwin Hyman: London. p. 294–304.

8. Geoffrey, H., *Environmental change and fire in the Owen Stanley Ranges, Papua New Guinea.* Quaternary Science Reviews, 2009. **28**(23–24): p. 2261–76.

9. Denham, T., et al., *Contiguous multi-proxy analyses (X-radiography, diatom, pollen, and microcharcoal) of Holocene archaeological features at Kuk Swamp, Upper Wahgi Valley, Papua New Guinea.* Geoarchaeology – An International Journal, 2009. **24**(6): p. 715–42.

10. Summerhayes, G.R., et al., *Human adaptation and plant use in highland New Guinea 49,000 to 44,000 years ago.* Science, 2010. **330**(6000): p. 78–81.

11. Denham, T., *Exploiting diversity: plant exploitation and occupation in the interior of New Guinea during the Pleistocene.* Archaeology in Oceania, 2007. **42**(2): p. 41–8.

12. Sutton, A., et al., *Archaeozoological records for the highlands of New Guinea: A review of current evidence.* Australian Archaeology, 2009. (69): p. 41–58.

13. Denham, T., M. Donohue, and S. Booth, *Horticultural experimentation in northern Australia reconsidered.* Antiquity, 2009. **83**(321): p. 634–48.

14. Harris, P.T., E.K. Baker, and A.R. Cole, *Late Quaternary sedimentation at the Fly River–Great Barrier Reef junction (NE Australia)*, in *Proceedings of the Seventh International Coral Reef Symposium*, 1992. University of Guam Press: Guam.

15. Flannery, T.F., *Throwim Way Leg: Tree-Kangaroos, Possums, and Penis Gourds—On the Track of Unknown Mammals in Wildest New Guinea*, 1998. Atlantic Monthly Press: New York.

16. Strathern, M., *Women in Between: Female Roles in a Male World: Mount Hagen, New Guinea*, 1972. Rowman & Littlefield: Lanham, MD.

17. Rotberg, R.I., *Corruption, Global Security, And World Order*, 2009. Brookings Institute Press: Washington DC.

18. Bulmer, S., *Pig bone from two archaeological sites in the New Guinea highlands.* The Journal of the Polynesian Society, 1966. **75**(4): p. 504–5.

19. Ponting, C., *A New Green History of the World: The Environment and the Collapse of Great Civilizations*, 2007. Vintage: London.

20. Fiedel, S., *Sudden deaths: The chronology of terminal Pleistocene megafaunal extinction*, in *American Megafaunal Extinctions at the End of the Pleistocene*, G. Haynes, editor, 2009. Springer: Netherlands. p. 21–37.

21. Dubelaar, C.N., *The Petroglyphs in the Guianas and the Adjacent Areas of Brazil and Venezuela: An Inventory with a Comprehensive Bibliography of South American and Antillean Petroglyphs*, 1986. University of California Press: Los Angeles.

22. Borrero, L.A., *The elusive evidence: The archeological record of the South American extinct megafauna*, in *American Megafaunal Extinctions at the End of the Pleistocene*, G. Haynes, editor, 2009. Springer: Netherlands. p. 145–68.

23. Martin, P.W., and H.E. Wright, eds. *Pleistocene Extinctions: The Search for a Cause*, 1967. Yale University Press: New Haven.

24. Hamilton, J.W., *Notes on Maori traditions of the moa*. Transactions of the Royal Society of New Zealand, 1874. **7**: p. 121–2.

25. McWethy, D.B., et al., *Rapid landscape transformation in South Island, New Zealand, following initial Polynesian settlement*. Proceedings of the National Academy of Sciences of the United States of America, 2010. **107**(50): p. 21343–8.

26. Aitchison, J.C., J.R. Ali, and A.M. Davis, *When and where did India and Asia collide?* Journal of Geophysical Research-Solid Earth, 2007. **112**(B5), article B05423.

27. Waters, M.R., et al., *Pre-Clovis mastodon hunting 13,800 years ago at the Manis Site, Washington*. Science, 2011. **334**(6054): p. 351–3.

28. Bement, L.C., and B.J. Carter, *Jake Bluff: Clovis bison hunting on the southern plains of North America*. American Antiquity, 2010. **75**(4): p. 907–33.

29. Bryan, A.L., et al., *An El Jobo mastodon kill at Taima-taima, Venezuela*. Science, 1978. **200**(4347): p. 1275–7.

30. Mussi, M., and P. Villa, *Single carcass of Mammuthus primigenius with lithic artifacts in the Upper Pleistocene of northern Italy*. Journal of Archaeological Science, 2008. **35**(9): p. 2606–13.

31. Grayson, D.K., *Deciphering North American Pleistocene extinctions*. Journal of Anthropological Research, 2007. **63**(2): p. 185–213.

32. Faith, J.T., *Late Pleistocene climate change, nutrient cycling, and the megafaunal extinctions in North America*. Quaternary Science Reviews, 2011. **30**(13–14): p. 1675–80.

33. Firestone, R.B., et al., *Evidence for an extraterrestrial impact 12,900 years ago that contributed to the megafaunal extinctions and the Younger Dryas cooling*. Proceedings of the National Academy of Sciences of the United States of America, 2007. **104**(41): p. 16016–21.

34. Perego, U.A., et al., *The initial peopling of the Americas: A growing number of founding mitochondrial genomes from Beringia*. Genome Research, 2010. **20**(9): p. 1174–9.

35. Fix, A.G., *Rapid deployment of the five founding Amerind mtDNA haplogroups via coastal and riverine colonization*. American Journal of Physical Anthropology, 2005. **128**: p. 430–6.

36. Nogues-Bravo, D., et al., *Climate change, humans, and the extinction of the woolly mammoth*. PLoS Biology, 2008. **6**(4): p. 685–92.

37. Mann, C.C., *1491: New Revelations of the Americas Before Columbus*, 2005. Knopf: New York.

38. Snow, D.R., *Microchronology and demographic evidence relating to the size of Pre-Columbian North-American Indian populations*. Science, 1995. **268**(5217): p. 1601–4.

39. Keeley, J.E., *Native American impacts on fire regimes of the California coastal ranges*. Journal of Biogeography, 2002. **29**(3): p. 303–20.

40. Haraldsson, H.V., and R. Olafsdottir, *A novel modelling approach for evaluating the preindustrial natural carrying capacity of human population in Iceland*. Science of the Total Environment, 2006. **372**(1): p. 109–19.

Chapter Eleven

1. Allman, J., *Evolving Brains*, 2000. W.H. Freeman: New York.

2. Wills, C., *The Runaway Brain: the Evolution of Human Uniqueness*, 1993. Basic Books: New York.

3. Stout, D., *Stone toolmaking and the evolution of human culture and cognition*. Philosophical Transactions of the Royal Society B-Biological Sciences, 2011. **366**(1567): p. 1050–9.

4. Wills, C., *Children of Prometheus: The Accelerating Pace of Human Evolution*, 1998. Perseus Books: Reading MA.

5. Aiello, L.C., and P. Wheeler, *The expensive-tissue hypothesis—The brain and the digestive-system in human and primate evolution*. Current Anthropology, 1995. **36**(2): p. 199–221.

6. Navarrete, A., C.P. van Schaik, and K. Isler, *Energetics and the evolution of human brain size*. Nature, 2011. **480**(7375): p. 91–3.

7. Barrickman, N.L., and M.J. Lin, *Encephalization, expensive tissues, and energetics: An examination of the relative costs of brain size in strepsirrhines*. American Journal of Physical Anthropology, 2010. **143**(4): p. 579–90.

8. Niven, J.E., and S.B. Laughlin, *Energy limitation as a selective pressure on the evolution of sensory systems*. Journal of Experimental Biology, 2008. **211**(11): p. 1792–804.

9. Griffin, D.R., *Sensory physiology and the orientation of animals*. American Scientist, 1953. **41**(2): p. 208–81.

10. Pepperberg, I., *Further evidence for addition and numerical competence by a Grey parrot (Psittacus erithacus)*. Animal Cognition. **15**(4): p. 711–17.

11. Larsen, C.C., et al., *Total number of cells in the human newborn telencephalic wall*. Neuroscience, 2006. **139**(3): p. 999–1003.

12. Sherwood, C.C., et al., *Evolution of increased glia-neuron ratios in the human frontal cortex*. Proceedings of the National Academy of Sciences of the United States of America, 2006. **103**(37): p. 13606–11.

13. Mink, J.W., R.J. Blumenschine, and D.B. Adams, *Ratio of central nervous-system to body metabolism in vertebrates—Its constancy and functional basis*. American Journal of Physiology, 1981. **241**(3): p. R203–12.

14. Diamond, M.C., et al., *Increases in cortical depth and glia numbers in rats subjected to enriched environment*. Journal of Comparative Neurology, 1966. **128**(1): p. 117.

15. Gelfo, F., et al., *Layer and regional effects of environmental enrichment on the pyramidal neuron morphology of the rat*. Neurobiology of Learning and Memory, 2009. **91**(4): p. 353–65.

16. Kolb, B., and I.Q. Whishaw, *Brain plasticity and behavior*. Annual Review of Psychology, 1998. **49**: p. 43–64.

17. Heisenberg, M., M. Heusipp, and C. Wanke, *Structural plasticity in the Drosophila brain*. Journal of Neuroscience, 1995. **15**: p. 1951–60.

18. Prata, D.P., et al., *Effect of D-amino acid oxidase activator (DAOA; G72) on brain function during verbal fluency*. Human Brain Mapping, 2012. **33**(1): p. 143–53.

19. Moore, C.I., and R. Cao, *The hemo-neural hypothesis: On the role of blood flow in information processing*. Journal of Neurophysiology, 2008. **99**(5): p. 2035–47.

20. Yuan, Q., et al., *Light-induced structural and functional plasticity in Drosophila larval visual system*. Science, 2011. **333**(6048): p. 1458–62.

21. Highnam, C.L., and K.M. Bleile, *Language in the cerebellum*. American Journal of Speech – Language Pathology, 2011. **20**(4): p. 337–47.

22. Dennis, M.Y., et al., *Evolution of human-specific neural SRGAP2 genes by incomplete segmental duplication*. Cell, 2012. **149**(4): p. 912–22.

23. Charrier, C.c., et al., *Inhibition of SRGAP2 function by its human-specific paralogs induces neoteny during spine maturation.* Cell, 2012. **149**(4): p. 923–35.

24. Nave, K.A., *Myelination and support of axonal integrity by glia.* Nature, 2010. **468**(7321): p. 244–52.

25. MacArthur, D.G., et al., *A systematic survey of loss-of-function variants in human protein-coding genes.* Science, 2012. **335**(6070): p. 823–8.

26. Blitzblau, R., E.K. Storer, and M.H. Jacob, *Dystrophin and utrophin isoforms are expressed in glia, but not neurons, of the avian parasympathetic ciliary ganglion.* Brain Research, 2008. **1218**: p. 21–34.

27. Ray, J.C.J., J.J. Tabor, and O.A. Igoshin, *Non-transcriptional regulatory processes shape transcriptional network dynamics.* Nature Reviews Microbiology, 2011. **9**(11): p. 817–28.

28. McClung, C.A., and E.J. Nestler, *Neuroplasticity mediated by altered gene expression.* Neuropsychopharmacology, 2007. **33**(1): p. 3–17.

29. Wang, H.-Y., et al., *Rate of evolution in brain-expressed genes in humans and other primates.* PLoS Biol, 2006. **5**(2): p. e13.

30. Somel, M., et al., *MicroRNA-driven developmental remodeling in the brain distinguishes humans from other primates.* PLoS Biol, 2011. **9**(12): p. e1001214.

31. Piao, X.H., et al., *G protein-coupled receptor-dependent development of human frontal cortex.* Science, 2004. **303**(5666): p. 2033–6.

32. Kamin, L.J., *The science and politics of I.Q.,* 1974. Halstead Press: New York.

33. Richardson, K., and D. Spears, eds. *Race and Intelligence: The Fallacies behind the Race-IQ Controversy,* 1972. Penguin Books: Baltimore MD.

34. Devlin, B., M. Daniels, and K. Roeder, *The heritability of IQ.* Nature, 1997. **388**(6641): p. 468–71.

35. Flynn, J.R., *Massive IQ gains in 14 nations: What IQ tests really measure.* Psychological Bulletin, 1987. **101**: p. 171–91.

36. Daley, T.C., et al., *IQ on the rise—The Flynn effect in rural Kenyan children.* Psychological Science, 2003. **14**(3): p. 215–19.

37. Nisbett, R.E., et al., *Intelligence: New findings and theoretical developments.* American Psychologist, 2012. **67**(2): p. 130–59.

38. Eppig, C., C.L. Fincher, and R. Thornhill, *Parasite prevalence and the worldwide distribution of cognitive ability.* Proceedings of the Royal Society B – Biological Sciences, 2010. **277**(1701): p. 3801–8.

39. Caspi, A., et al., *Moderation of breastfeeding effects on the IQ by genetic variation in fatty acid metabolism.* Proceedings of the National Academy of Sciences of the United States of America, 2007. **104**(47): p. 18860–5.

40. Butcher, L.M., et al., *Genome-wide quantitative trait locus association scan of general cognitive ability using pooled DNA and 500K single nucleotide polymorphism microarrays.* Genes, Brain and Behavior, 2008. **7**(4): p. 435–46.

41. Park, K.I., et al., *Global gene and cell replacement strategies via stem cells.* Gene Therapy, 2002. **9**(10): p. 613–24.

42. Gardner, H., *Intelligence Reframed: Multiple Intelligences for the 21st Century,* 1999. Basic Books: New York.

43. Gravel, S., et al., *Demographic history and rare allele sharing among human populations.* Proceedings of the National Academy of Sciences of the United States of America, 2011. **108**(29): p. 11983–8.

44. Martin, N.W., et al., *Educational attainment: A genome wide association study in 9538 Australians.* PLoS ONE, 2011. **6**(6), e20128.

45. Luciano, M., et al., *Whole genome association scan for genetic polymorphisms influencing information processing speed.* Biological Psychology, 2011. **86**(3): p. 193–202.

46. Crisci, J.L., et al., *On characterizing adaptive events unique to modern humans.* Genome Biology and Evolution, 2011. **3**: p. 791–8.

47. Hawks, J., et al., *Recent acceleration of human adaptive evolution.* Proceedings of the National Academy of Sciences, 2007. **104**(52): p. 20753–8.

48. Green, R.E., et al., *A draft sequence of the Neandertal genome.* Science, 2010. **328**(5979): p. 710–22.

Chapter Twelve

1. Catholic Relief Services, *Nepal Dam Break—Floods in Bihar, India.* 2008. Available from: http://reliefweb.int/node/279142.

2. Ryback, T.W., *The U.N. Happiness Project*, in *New York Times*, 2012: New York.

3. Dorji, G., *Not Yet Happy More Than Happy.* Kuensel Online, 2012.

4. Helliwell, J., R. Layard, and J. Sachs, *World Happiness Report*, 2012.

5. Fujii, R., and H. Sakai, *Paleoclimatic changes during the last 2.5 myr recorded in the Kathmandu Basin, Central Nepal Himalayas.* Journal of Asian Earth Sciences, 2002. **20**(3): p. 255–66.

6. Brown, L.R., and E. Eckholm, *By Bread Alone*, 1974. Praeger: New York.

7. Ale, S.B., P. Yonzon, and K. Thapa, *Recovery of snow leopard Uncia uncia in Sagarmatha (Mount Everest) National Park, Nepal.* Oryx, 2007. **41**(01): p. 89–92.

8. Himalayan News Service, *Rampant poaching in Rara National Park*, in *The Himalayan*, 2012: Kathamandu, Nepal. http://www.thehimalayantimes.com/fullNews.php?headline=Rampant+poaching+in+Rara+National+Park+&NewsID=325370.

9. Durrell, G.M., *Three Singles to Adventure*, 1954. Rupert Hart-Davis: London.

10. Shackley, M., *Designating a protected area at Karanambu ranch, Rupununi Savannah, Guyana: Resource management and indigenous communities.* Ambio, 1998. **27**(3): p. 207–10.

11. McTurk, D., and L. Spelman, *Hand-rearing and rehabilitation of orphaned wild giant otters, Pteronura brasiliensis, on the Rupununi River, Guyana, South America.* Zoo Biology, 2005. **24**(2): p. 153–67.

12. Rupununi Learners, *Black caiman field study.* 2012. Available from: http://www.rupununilearners.org/biological_research/black_caiman.htm.

L'Envoi

1. Kalmar, A., and D.J. Currie, *The completeness of the continental fossil record and its impact on patterns of diversification.* Paleobiology, 2010. **36**(1): p. 51–60.

2. Alroy, J., et al., *Phanerozoic trends in the global diversity of marine invertebrates.* Science, 2008. **321**(5885): p. 97–100.

INDEX